T0234106

External Labeling

Fundamental Concepts and Algorithmic Techniques

Synthesis Lectures on Visualization

Editors
Niklas Elmqvist, *University of Maryland*
David S. Ebert, *University of Oklahoma*

Synthesis Lectures on Visualization publishes 50- to 100-page publications on topics pertaining to scientific visualization, information visualization, and visual analytics. Potential topics include, but are not limited to: scientific, information, and medical visualization; visual analytics, applications of visualization and analysis; mathematical foundations of visualization and analytics; interaction, cognition, and perception related to visualization and analytics; data integration, analysis, and visualization; new applications of visualization and analysis; knowledge discovery management and representation; systems, and evaluation; distributed and collaborative visualization and analysis.

External Labeling: Fundamental Concepts and Algorithmic Techniques
Michael A. Bekos, Benjamin Niedermann, and Martin Nöllenburg
2021

Visual Analysis of Multilayer Networks
Fintan McGee, Benjamin Renoust, Daniel Archambault, Mohammad Ghoniem, Andreas Kerren, Bruno Pinaud, Margit Pohl, Benoît Otjacques, Guy Melançon, and Tatiana von Landesberger
2021

Adaptive and Personalized Visualization
Alvitta Ottley
2020

Adaptive and Personalized Visualization
Alvitta Ottley
2020

Diversity in Visualization
Ron Metoyer and Kelly Gaither
2019

User-Centered Evaluation of Visual Analytics
Jean Scholtz
2017

Interactive GPU-based Visualization of Large Dynamic Particle Data
Martin Falk, Sebastian Grottel, Michael Krone, and Guido Reina
2016

Semantic Interaction for Visual Analytics: Inferring Analytical Reasoning for Model
Steering
Alexander Endert
2016

Design of Visualizations for Human-Information Interaction: A Pattern-Based
Framework
Kamran Sedig and Paul Parsons
2016

Image-Based Visualization: Interactive Multidimensional Data Exploration
Christophe Hurter
2015

Interaction for Visualization
Christian Tominski
2015

Data Representations, Transformations, and Statistics for Visual Reasoning
Ross Maciejewski
2011

A Guide to Visual Multi-Level Interface Design From Synthesis of Empirical Study
Evidence
Heidi Lam and Tamara Munzner
2010

External Labeling: Fundamental Concepts and Algorithmic Techniques

Michael A. Bekos, Benjamin Niedermann, and Martin Nöllenburg

ISBN: 978-3-031-01481-9   paperback
ISBN: 978-3-031-02609-6   ebook
ISBN: 978-3-031-00353-0   hardcover

DOI 10.1007/978-3-031-02609-6

A Publication in the Springer series
SYNTHESIS LECTURES ON VISUALIZATION

Lecture #13
Series Editors: Niklas Elmqvist, *University of Maryland*
                David S. Ebert, *University of Oklahoma*
Series ISSN
Print 2159-516X   Electronic 2159-5178

External Labeling

Fundamental Concepts and Algorithmic Techniques

Michael A. Bekos
Universität Tübingen

Benjamin Niedermann
Universität Bonn

Martin Nöllenburg
TU Wien

SYNTHESIS LECTURES ON VISUALIZATION #13

ABSTRACT

This book focuses on techniques for automating the procedure of creating external labelings, also known as *callout labelings*. In this labeling type, the features within an illustration are connected by thin leader lines (called *leaders*) with their labels, which are placed in the empty space surrounding the image.

In general, textual labels describing graphical features in maps, technical illustrations (such as assembly instructions or cutaway illustrations), or anatomy drawings are an important aspect of visualization that convey information on the objects of the visualization and help the reader understand what is being displayed.

Most labeling techniques can be classified into two main categories depending on the "distance" of the labels to their associated features. *Internal labels* are placed inside or in the direct neighborhood of features, while *external labels*, which form the topic of this book, are placed in the margins outside the illustration, where they do not occlude the illustration itself. Both approaches form well-studied topics in diverse areas of computer science with several important milestones.

The goal of this book is twofold. The first is to serve as an entry point for the interested reader who wants to get familiar with the basic concepts of external labeling, as it introduces a unified and extensible taxonomy of labeling models suitable for a wide range of applications. The second is to serve as a point of reference for more experienced people in the field, as it brings forth a comprehensive overview of a wide range of approaches to produce external labelings that are efficient either in terms of different algorithmic optimization criteria or in terms of their usability in specific application domains. The book mostly concentrates on algorithmic aspects of external labeling, but it also presents various visual aspects that affect the aesthetic quality and usability of external labeling.

KEYWORDS

- external labeling: boundary, contour, excentric, static, dynamic

- points, features, leaders, labels, callouts

- taxonomy, quality metrics, optimization, algorithmic techniques

Contents

Bibliography i

Preface ix

Acknowledgments xi

Figure Credits xiii

1   Introduction 1
   1.1   External Labeling in Applications 6
   1.2   Book Structure 8
   1.3   How to Read This Book 9

2   A Unified Taxonomy 11
   2.1   Terminology and Concepts 11
   2.2   Distinctive Features 15
       2.2.1 Admissible Positions of Labels 15
       2.2.2 Leader Type 17
       2.2.3 Static or Dynamic Labeling 18
   2.3   Optimization Problem 19

3   Visual Aspects of External Labeling 23
   3.1   Style 23
   3.2   Placement 25
   3.3   Empirical Studies 28

4   Labeling Techniques 31
   4.1   Non-Exact Algorithms 31
       4.1.1 Greedy Algorithms 31
       4.1.2 Force-Based Algorithms 33
       4.1.3 Miscellaneous Techniques 34
   4.2   Exact Algorithms 35

        4.2.1 Dynamic Programming 35

        4.2.2 Weighted Matching 40

        4.2.3 Scheduling 42

        4.2.4 Plane Sweep 43

        4.2.5 Mathematical Programming 45

  4.3   Complexity Results 47

  4.4   Guidelines 48

5   External Labelings with Straight-Line Leaders 51

  5.1   Overview 51

  5.2   Detailed Discussion 57

        5.2.1 Contour Labeling 57

        5.2.2 Free Label Placement 62

        5.2.3 Non-Strict External Labeling 63

  5.3   Guidelines 65

6   External Labelings with Polyline Leaders 69

  6.1   Overview 69

  6.2   Detailed Discussion 77

        6.2.1 po-Leaders 77

        6.2.2 opo-Leaders 83

        6.2.3 do- and pd-Leaders 89

        6.2.4 Other Polyline Leaders 91

  6.3   Curved Leaders 91

  6.4   Guidelines 92

7   Conclusions and Outlook 95

8

Correction to: Visual Aspects of External Labeling C1

Bibliography 101

Authors' Biographies 111

Index 113

Preface

This book is the final outcome of continuing discussions among the three authors regarding the growing body of literature on the topic of external labeling that we were facing when discussing related work in several of our own articles on the topic. Accordingly, in March 2018, we started to intensify our discussions and collected and structured the related literature with the first goal to write a state-of-the-art report for the EuroVis conference in Lisbon in 2019. Shortly after presenting our survey at the conference, we were contacted by Niklas Elmqvist and David Ebert, editors of this *Synthesis Lectures in Visualization* series. They invited us to extend our compact survey paper into a comprehensive book and the result of this project is in your hands now.

The field of external labeling is mature enough to warrant a comprehensive handbook as a resource both for researchers with first experience in external labeling who want to broaden their knowledge, as well as for new students and researchers coming across problems in their research that relate to external labeling. However, we also believe that practitioners and domain experts who are in need of finding and implementing suitable labeling algorithms for their instances at hand or who want to obtain a deeper understanding of the strengths and weaknesses of their currently used methods will find value in this book and its introduction to the technical background of the topic.

As a team of three researchers with a background on formal, algorithmic methods in graph drawing, computational geometry, and information visualization, we have worked ourselves on many external labeling problems, both from theoretical and practical perspectives. After years of research experience it turned out that there is a small set of algorithm design techniques, which can be used to solve a large number of external labeling problems. Thus, one goal of this book is to summarize and explain these techniques to readers with different backgrounds in computer science and related disciplines.

Moreover, we observed that there is a multitude of labeling models with various important parameters, but no commonly used taxonomy guiding experts and novices alike through the existing state of the art. This is due not least because external labeling is studied in many different fields such as algorithm design, information visualization, computer graphics, or virtual/augmented reality, all with their own approaches to the respective problems—from the mathematical curiosity of basic research to the practical needs of creating readable visualizations. A second goal of the book is thus to unify the diverse labeling models and provide a common taxonomy, which facilitates classifying new research results.

A third part of the book covers the existing state of the art in a well-structured way, both in a compact tabular form—where each method is described according to a set of important parameters—as well as in a more detailed description of the respective results. Finally, we provide a collection of ten research challenges in external labeling to be seen as opportunities for interdisciplinary research collaborations in the coming years.

We hope you enjoy reading this book and find it useful for your own work.

Michael A. Bekos, Benjamin Niedermann, and Martin Nöllenburg
June 2021

Acknowledgments

This book is the result of several years of research experience on external labeling. That said, we deem it important to thank several colleagues and co-authors with whom we have closely collaborated over the years. In close relation to this book, we would also like to thank Denis Kalkofen for fruitful discussions and for exchanging useful ideas with us. The contribution of all reviewers of this book (both named and anonymous) should also be acknowledged, since their insightful comments and suggestions helped in significantly improving both the content and the presentation of this book. Last, but not least, special thanks go to editors Niklas Elmqvist and David Ebert, who invited us to extend a preliminary version of this book into a contribution to the *Synthesis Lectures on Visualization* series, and to Diane Cerra and her team at Morgan & Claypool, who provided us with useful support and helped us accomplish this project.

Michael A. Bekos, Benjamin Niedermann, and Martin Nöllenburg
June 2021

Figure Credits

- Figure 1.1(a). Map tiles by Stamen Design, under CC BY 3.0. Data by Open-StreetMap, under ODbL.

- Figure 1.1(b): https://commons.wikimedia.org/wiki/File:Mercosur-map-fr.svg. Public domain.

- Figure 1.2(a): Atlas of applied (topographical) human anatomy for students and practitioners. K. H. v. Bardeleben, E. Haeckel, Rebman Company, 1906. Public domain.

- Figure 1.2(b), Figure 5.9: www2.geoinfo.uni-bonn.de/html/fisheyelabeling. Map tiles by Stamen Design, under CC BY 3.0. Map data by OpenStreetMap, under ODbL.

- Figure 1.2(c), Figure 6.9: courtesy of DW-TV.

- Figure 1.2(d): https://ourworldindata.org/coronavirus, under CC BY 3.0.

- Figure 1.3(a): https://commons.wikimedia.org/wiki/File:Culex_restuans_larva_diagram_en.svg. Public domain.

- Figure 1.3(b): https://commons.wikimedia.org/wiki/File:Bicycle_diagram-en.svg, Al2, under CC BY A 3.0.

- Figure 1.3(c): https://commons.wikimedia.org/wiki/File:Plant_cell_structure-en.svg. Public domain.

- Figure 1.3(d): https://commons.wikimedia.org/wiki/File:HubbleExploded_edit_1.svg, under CC BY SA 3.0, GFDL.

- Figure 1.5: https://www.peakfinder.org/de/mobile. Image courtesy of Fabio Soldati, PeakFinder GmbH.

- Figure 2.8(a): www2.geoinfo.uni-bonn.de/html/radiallabeling. Map tiles by Stamen Design, under CC BY 3.0. Map data by OpenStreetMap, under ODbL.

- Figure 2.8(b): Figure courtesy of Ladislav Čmolík; created with the method presented in L. Čmolík and J. Bittner. Real-time external labeling of ghosted views. *IEEE Transactions on Visualization and Computer Graphics*, 2018. doi:10.1109/TVCG.2018.2833479.

CHAPTER 1

Introduction

Depicting complex information in a clean, human-friendly, and easily understandable manner is a task, with which not just professional illustrators and graphic designers are faced in their everyday work. Journalists, scientists, engineers, and many other professions, as well as ordinary citizens in their private life, may more or less frequently come across the task to explain and describe sketches, detailed illustrations, pictures, diagrams, or photographs, to just name a few. But even if a suitable image has been created in a first step, let's say an assembly plan of a machine, a sketch of a human cell, or a group photo of a wedding party, the task is generally not yet done. For the audience of the image to understand the full details, critical parts of the image need to be annotated with a name or a short description: the part identifiers in the assembly plan, the cell compartments and organelles in the human cell, or the names of the guests on the photo.

An area with long-standing experience in annotating graphical information is cartography. Maps are extremely detailed and sophisticated displays of spatial information, from small-scale land register maps to hiking maps, from political maps to entire globes. Places and areas on such maps are typically labeled by placing their names in the direct vicinity of each such feature; see Figure 1.1(a) for an example. In this case, the association between a feature and its label is derived by their proximity. Such a label next to its feature is also called an *internal label*. Cartographers are facing many issues when creating and labeling maps. Usually, the number of features of interest is much higher than there is space for their labels to be placed. Hence, a subset of the most relevant features must be selected, for which it is possible in a second step to place all names. For instance, in Figure 1.1(a) some countries in Europe did not receive a label. This obviously results in a loss of certain information, which is not always acceptable.

More precisely, with images that have already abundant graphical detail, placing such object labels as in a cartographic map is not as straightforward as it might seem at first sight. Simply putting labels on top or right next to all features of interest quickly leads to overplotting and occlusion of relevant graphical information, even if the names themselves would be free of overlap. Even worse, in areas of high feature density, there might simply not be enough space to place all names without text overlap. The most common and natural solution to this problem is to place the names outside of the image or in areas, where no important details are occluded, see Figure 1.1(b) for a map with some displaced labels. To associate each name with its image feature and vice versa, a connecting line, a so-called *leader*, is needed to create a unique and unambiguous link. Figure 1.2(a) shows a second example: a hand-drawn and labeled illustration of the fourth ventricle, a part of the human brain, from a historic atlas of human anatomy. This

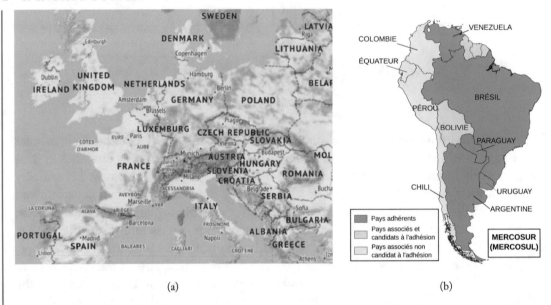

(a)   (b)

Figure 1.1: Examples for label placement in cartographic maps. (a) Internal labeling of countries (areas) and cities (points) in Europe. (b) Mixed labeling of countries in South America. While some labels are external, other labels are internal.

illustration is so rich in detail that a leader-based labeling was seen as the best way of conveying all the necessary information to the medical students. Such a labeling style using leaders to connect features of interest with their labels is known as *external labeling* or *callout labeling* in the literature.

External labelings come in many different shapes. Some examples are shown in Figures 1.2 and 1.3. An important property of external labeling is the separation of an image region from a labeling region, where all features are contained in the image region while all labels are placed in the labeling region. The separation curve between the two regions can be explicit, such as a bounding box or silhouette (Figure 1.2(a)), or more implicit and subtle, such as a circular focus region (Figure 1.2(b)) or an invisible vertical boundary (Figures 1.2(c) and 1.2(d)). There is not always a clear-cut division between internal and external labeling. Some real-world examples use a mixed labeling model with internal labels where there is sufficient space and external labels otherwise; Figure 1.1(b) shows an example of such a mixed labeling. Sometimes features and their labels are connected by leaders, but there is no designated labeling area and callouts are placed in a free location as close to its feature as possible, e.g., see Paraguay in Figure 1.1(b) .

There are basically two different ways of obtaining an external labeling for a given image: it can be created by a human who decides the positioning of labels and the shape of the leader, either manually on paper or using drawing tools in image processing software, or it can be created

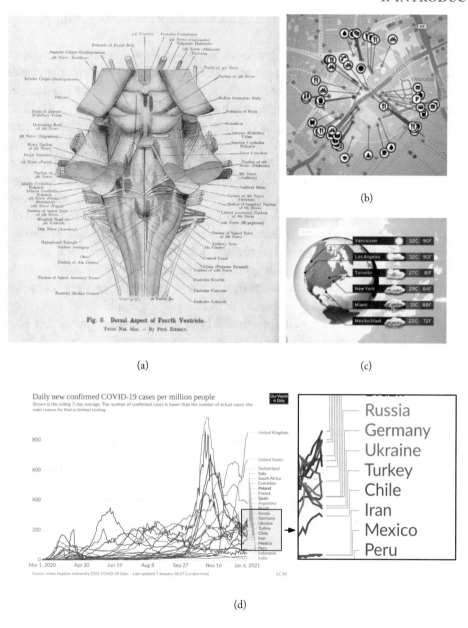

Figure 1.2: Different types of external labelings. (a) Contour labeling: anatomical illustration from *Atlas of applied (topographical) human anatomy for students and practitioners. K. H. v. Bardeleben, E. Haeckel*, Rebman Company, 1906. (b) Excentric labeling: fish-eye labeling produced with the approach by Niedermann and Haunert [79]. Boundary labeling: (c) weather map of the U.S. (d) Boundary labeling: the daily new confirmed COVID-19 cases per million people.

automatically by a labeling algorithm optimizing the placement and routing of the leaders. Both approaches have their pros and cons.

Skilled and experienced human professionals are generally well aware of the overall aesthetics and readability of their labeled images and can create balanced and visually pleasing results. While intuition may guide labeling experts to very good results most of the time, solving complex label placement tasks in a satisfying way involves exploration of a huge solution space and can become a very difficult task if one lacks experience or one faces an instance with a large number of features to be labeled. Hence, it does not come as a surprise that label placement by a human professional is both time-intensive and costly for the commissioner of the order.

On the other hand, algorithmic solutions promise to be faster by orders of magnitude than a human designer. In some use-cases such as interactive visualizations or virtual and augmented reality, automatically placing labels is the only option as the scene to be labeled changes dynamically. Other examples, such as collaborative document editing, produce primarily static labelings, but again the labels must be placed conveniently by an algorithm. Such automatically produced solutions, however, can be only as good as the specification of the admitted labelings and the mathematical model of the solution quality. Further, depending on the chosen algorithmic technique and the computational complexity of the underlying combinatorial and geometric optimization problem, the scalability and practicability of the automated approaches can differ dramatically. This book focuses on algorithmic approaches for external labeling and sets out to give a systematic overview and introduction to the variety of external labeling algorithms developed and presented in multiple subdisciplines of computer science, such as computational geometry, graph drawing, information visualization, computer graphics, and virtual/augmented reality, as well as in related field such as cartography and GIScience.

Over the last 20 years various aspects of the external labeling problem have been considered in the literature—both from a theoretical and from a practical point of view. The range of topics is so wide that we use the term "external labeling" mostly as an umbrella term, which covers several different labeling approaches, such as *contour labeling* [3, 80] (see Figure 1.2(a)), *boundary labeling* [16, 98] (see Figure 1.2(c) and Figure 1.2(d)), and *excentric* or *focus–region labeling* [25, 37] (see Figure 1.2(b)).

These terms refer to where the external labels can be placed with respect to the illustrated image, either tracing the image contour, on the axis-parallel sides of a bounding box, or along the circle of an interactive magnifying lens. Labels can come in groups (see Figure 1.3(b)) or are uniformly placed around the illustration (see Figure 1.3(d)). Our taxonomy in Chapter 2 will provide a detailed introduction to the different types of external labeling. In addition to the positioning of the labels, a second important characteristic of an external labeling is the shape of the leaders. While straight-line leaders (see Figures 1.2(a), 1.2(b), 1.3(b), or 1.3(c)) provide the shortest and most natural link between a feature and its label, they are not always the first choice of a designer as having too many distinct leader slopes can also lead to visual clutter. Thus, alternatively, many external labelings also use leaders consisting of polylines with one or

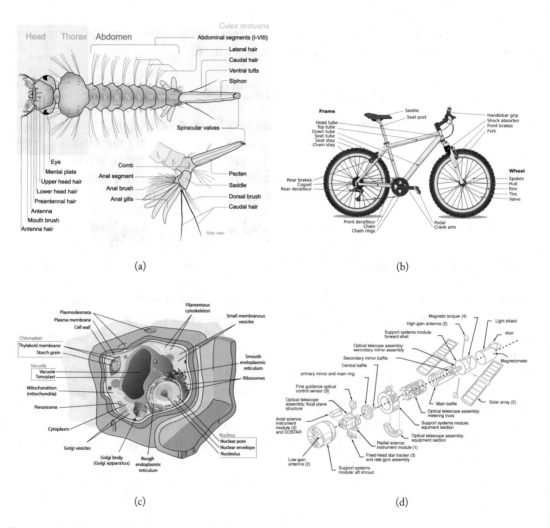

Figure 1.3: External labelings using different types of leaders. (a) Anatomy of a mosquito: straight-line leaders and leaders with bends. (b) Technical parts of a bicycle: straight-line leaders with grouped labels. (c) Structure of a plant cell: straight-line leaders with labels naming multiple features. (d) Explosion diagram of the Hubble telescope: leaders with bends.

more bends and often restricted to axis-aligned or diagonal segment slopes (see Figures 1.2(c), 1.2(d), 1.3(a), or 1.3(d)). Such leaders can create a more organized and schematic appearance, and can be used as a visual contrast to less structured and more messy background illustrations. In addition to the shape of the leaders, there can also be many-to-one leaders (or hyperleaders) that link a single label to multiple identical features (see Figure 1.3(c)). Again, the taxonomy in Chapter 2 will define the various leader types in more detail.

1.1   EXTERNAL LABELING IN APPLICATIONS

We distinguish between static and dynamic applications for external labeling. In static applications the setting is fixed, i.e., the contour of the image as well as the positions of the features to be labeled are known in advance. In contrast, in dynamic applications the underlying figure changes over time either by a pre-defined motion or by the interaction of a user exploring the details of a visualization or navigating in virtual or augmented reality spaces.

Static applications   Standard examples for external labeling in static applications are atlases of (human) anatomy or visual dictionaries; see Figures 1.2(a) and 1.3(b). They often come as printed books and comprise a large collection of highly detailed illustrations. Typically, each illustration is placed on its own page. Further, the labels naming the features are placed in the margins of the illustration in order to avoid unnecessary occlusions with the illustration. As a sloppy labeling can easily spoil the appearance of an entire page, the designers of such books invest much time into the process of placing labels. A good placement requires that the labels as well as the leaders connecting the labels with their features blend in with the illustration and the content of the pages. This particularly includes a good spacing between labels, short leaders as well as a composedly visual pattern of the entire labeling. Moreover, the labelings of different illustrations should have the same style such that the entire book appears to be made in one piece. As creating such labelings is complicated and requires the implementation of aesthetics principles, they are mostly created by hand. So far, there is only little tool support for designers, but automatic approaches that meet the high quality standards are still missing.

Another example for static applications is collaborative text editing using word processors such as Microsoft Word, Pages by Apple, or LaTeX; see Figure 1.4. In these applications the users can annotate written text with additional comments that are placed in the margins of the text. Accordingly, there is only little space for few comments per page. Further, the leaders of the labels should not cross the text but run in between the text lines. This particularly requires that the routing of the leaders is also done in the margins of the text further limiting the space for the placement of labels. While for professional books the placement can still be done by hand or in a semi-automatic way, for annotating texts fast and completely automatized approaches are required.

Finally, we note that also in diagrams external label placement is used to name the diagram's features. As diagrams often show curves, points and other geometric shapes it is of high

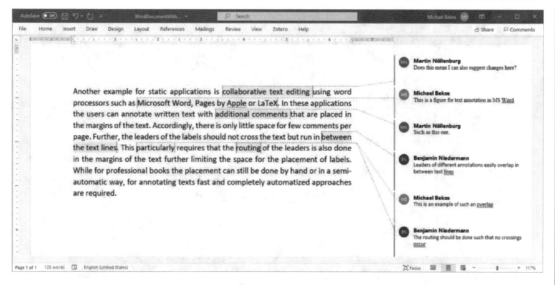

Figure 1.4: Collaborative text editing using Microsoft Word.

importance that not only the labels, but also the leaders, do not interfere with these geometric elements to avoid confusion. One placement strategy is therefore to place the labels and route the leaders entirely in the outside of the diagram; see Figure 1.2(d).

Dynamic applications   In dynamic applications not only a single illustration, but an entire sequence of illustrations, is annotated with labels. Examples are movie sequences as well as animations. In these applications it is important that the mental map is preserved, which requires that the labelings are temporally coherent from frame to frame. Hence, further requirements are enforced, in addition to the design principles for static applications. Typically, labels may not jump or flicker during the animation. In order to achieve this, approaches allow crossings between leaders or soften the restriction that labels are entirely placed in the exterior of the illustration. For these applications the change of the illustration is known in advance such that the labeling can be created in a pre-processing phase. This allows users to apply time-consuming approaches that strongly optimize the labelings with respect to given design principles. However, at the latest when the change of the illustration is done by the user interactively, approaches that create labeling in real time become necessary. Applications are information systems showing 3D models, immersive environments, video games, and augmented reality. For instance, Figure 1.5 shows a screenshot of an app for real-time labeling of mountain peaks on a live camera feed using vertical leaders and slanted labels.

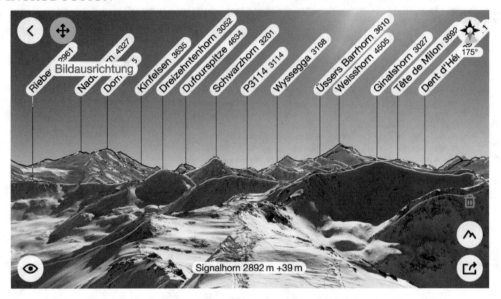

Figure 1.5: Screenshot of the PeakFinder augmented reality smartphone app for labeling mountain peaks.

1.2 BOOK STRUCTURE

Due to the multi-disciplinary nature of the research and application areas interested in external labeling, this book first establishes a common taxonomy to describe and specify different external labeling settings, visual quality criteria, and formal optimization models. This taxonomy is presented in Chapter 2. It serves both as an introduction to the topic for researchers who are newly entering the area and as a common basis for those with some previous experience who want to extend their scope. Similarly, it shall help practitioners and domain users who are interested in obtaining more background knowledge about the different algorithms for external labeling. By understanding their mechanics, strengths, and weaknesses, the book may assist them in selecting the most appropriate labeling model and corresponding algorithms for their particular data visualization and labeling tasks.

Even though the focus of this book is primarily on algorithmic techniques for external labeling, Chapter 3 gives a detailed summary of the most important visual aspects that affect the aesthetic quality and usability of external labelings. Besides the style of the labeling and the actual placement of the labels, we discuss several important criteria that must be taken into account when deciding on the specific type of external labeling to use, which may also depend on the needs of the particular application domain.

In Chapter 4, we identify and discuss in detail a collection of eight algorithmic techniques that are frequently used in the literature to create external labelings. We distinguish between exact and non-exact algorithmic techniques. The former solve instances of external labeling problems and provide a formal optimality guarantee for the solution, while the latter refrain from such guarantees in favor of higher efficiency. In both cases, we discuss properties and limitations as a guideline for researchers and practitioners interested in external labeling.

Chapters 5 and 6 provide a comprehensive literature overview of the most important research contributions in external labeling ranging from theoretical and algorithmic results to practical results in visual computing, including some empirical studies. Both chapters conclude with guidelines for selecting a labeling model and a suitable labeling algorithm. The dichotomy of the two chapters is based on the shape of the leaders that realize the connections between the features and their labels.

- Straight-line leaders form the simplest way to establish the visual association between features and their external labels, as they can be easily traced by the reader (Chapter 5).

- Polyline leaders form a natural generalization of straight-line leaders by supporting bends (Chapter 6), usually just one or two per leader to facilitate readability.

We conclude in Chapter 7 with a discussion of several challenges in external labeling from various technical perspectives, which provide ample opportunities for future work. These challenges range from algorithmic problems and less explored labeling models (e.g., combining external and internal labeling), to perceptual questions and studies of human comprehensibility of the labeled data.

1.3   HOW TO READ THIS BOOK

As already mentioned, the book is organized into three main parts. The first part consists of Chapters 2 and 3, and introduces basic concepts, various labeling parameters, and different visual aspects of external labeling, which, in our opinion, can help interested researchers and practitioners to enter the area and gain basic background knowledge. We expect that this introduction will also help researcher familiar with other labeling approaches to further deepen their understanding on the various design options that may result in different labeling results.

The second part consists of Chapter 4, which covers more advanced algorithmic techniques that are featured prominently in current research, such as specific heuristics, greedy algorithms, dynamic programming, and plane sweep. This part of the book mostly addresses readers with algorithmic background. Our goal is to help the algorithm designer by providing several algorithmic techniques that are likely to be helpful in practice, without entering into their complexity in greater detail.

The third part, consisting of Chapters 5–7, introduces the reader to the current state-of-the-art research by providing detailed references to the original publications, grouped by different labeling aspects. It further provides a detailed discussion on future challenges that are of importance for external labeling. According to our opinion, this part addresses mostly domain experts, who are interested in finding suitable labeling algorithms and understanding the strengths and weaknesses of the different methods. At the same time, this part may also serve as a structured entry point for designers of labeling algorithms into the state-of-the-art literature in external labeling.

CHAPTER 2

A Unified Taxonomy

In this chapter, we present a unified and extensible taxonomy of different labeling models proposed in the literature. At the same time, it serves as an entry point for the interested reader who wants to get familiar with the basic concepts of external labeling. First, we formally introduce the most important terminology and concepts on external labeling (Section 2.1). Figure 2.1 gives an overview of the most relevant terms used in this book and shall serve the reader as an overview of the terminology. Afterward, we present distinctive features that one may use to characterize and unify the different models for external labeling found in the literature (Section 2.2). Finally, we formally define a generic optimization problem that underlies approaches for creating external labelings automatically (Section 2.3).

2.1 TERMINOLOGY AND CONCEPTS

In general, a figure with external labeling can be decomposed into three layers; see Figure 2.2. Roughly speaking, the bottom layer defines the image area containing the illustration and the labeling area. The middle layer contains the features to be labeled. The top layer finally contains the annotations for the features. In the following, we discuss the different layers in greater detail.

Background layer The bottom layer partitions the available *drawing area D* along a simple, closed curve C_R into an *image region R* and a *labeling region $A = D \setminus R$*; we call this layer the *background layer*. The image region contains the unlabeled illustration, while the labeling region defines the empty space surrounding the image, which is reserved for placing the annotations. For instance, the image region could be a bounding box or a convex hull of the illustration.

Feature layer The middle layer defines a set F of pairwise disjoint features of interest to be labeled; we call this layer the *feature layer*. Each feature is a small sub-region of R, which can be a point, a curve or an area. Further, each feature has some attached information, e.g., its name, an icon, or a short textual description, which will later be used as the content of its label.

Labeling layer The top layer finally adds the labels and leaders to the figure—we call this layer the *labeling layer*; refer also to Figure 2.3 for a detailed example. Geometrically, we define the *label* as the axis-aligned bounding box of the feature's attached information. We denote the set of all labels by L. Typically, each feature has its own label, but in some special cases multiple features can share the same label or one feature can have several labels. More precisely, the label ℓ of a feature f is placed within the labeling region A and connected to f by a simple curve λ; we

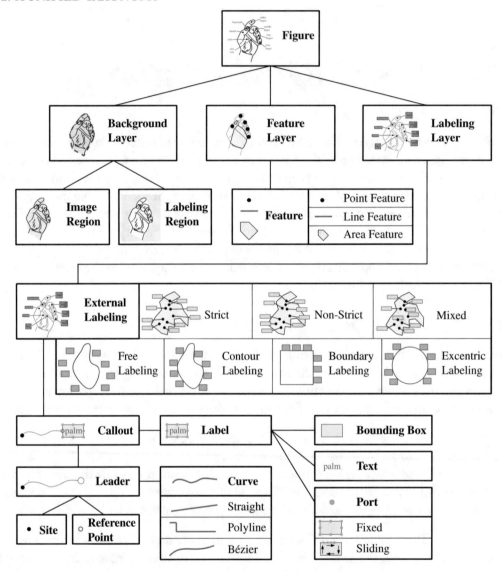

Figure 2.1: Composition tree for the terminology used throughout this book.

call λ the *leader* of ℓ and f. One end point of λ, the *site* $\sigma(\lambda)$ (sometimes also called *anchor*), is a point contained in f. The other end point of λ, the so-called *reference point* $a(\lambda)$, is a point on the boundary of ℓ.

We distinguish different ways of placing sites, reference points of leaders and labels. If the site σ can be any point of f, then σ is a *free* site. If σ is restricted to be any point of a pre-defined

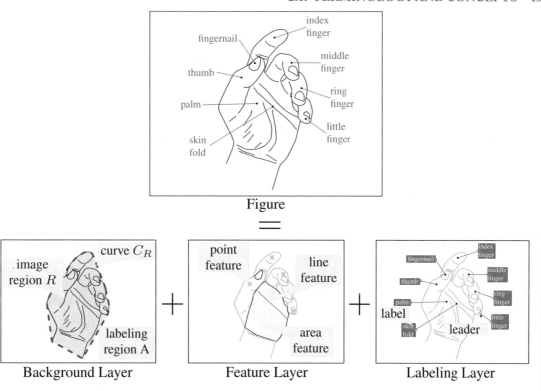

Figure 2.2: Layer decomposition. The top figure is decomposed into the background, feature, and labeling layer.

curve in f, then σ is a *sliding* site. Finally, if σ is restricted to be any point of a finite set of points in f, then σ is a *fixed* site. Note that the site of a point feature is always fixed, in fact the site and the feature coincide. Similarly, if the reference point a can be any point in A, then a is a *free* reference point. If a is restricted to be any point of a curve C_A in A, then a is a *sliding* reference point. Finally, if a is restricted to be any point of a finite set P_A of points in A, then a is a *fixed* reference point.

The most common cases are the use of sliding or fixed reference points, where the curve C_A or the set P_A either coincide with C_R or are offset by a small distance from C_R. The case of free reference points is less relevant, because labels should generally be placed close to the image region. Recall that the reference point a is a point on the boundary of the label ℓ. From the perspective of ℓ we call this attachment point between the leader λ and ℓ the *port* $\pi(\ell)$ of ℓ, which implies $a(\lambda) = \pi(\ell)$. If the port $\pi(\ell)$ can be any point on the boundary of ℓ, then $\pi(\ell)$ is a *sliding* port. If $\pi(\ell)$ is restricted to be any point of a finite set of points on the boundary of ℓ, then $\pi(\ell)$ is a *fixed* port. The most common fixed-port model restricts the ports to the corners

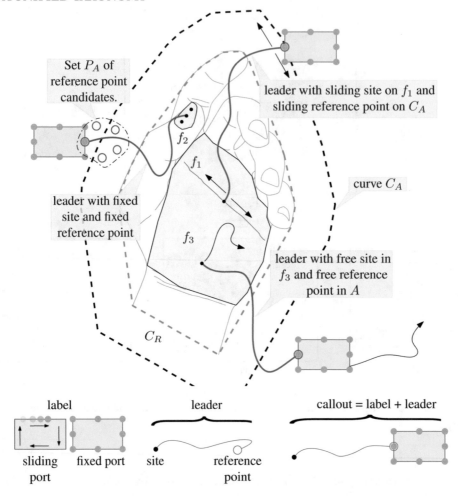

Set P_A of reference point candidates.

leader with sliding site on f_1 and sliding reference point on C_A

curve C_A

leader with fixed site and fixed reference point

leader with free site in f_3 and free reference point in A

C_R

label

sliding port   fixed port

leader

site   reference point

callout = label + leader

Figure 2.3: Illustration of terminology used throughout this book.

or midpoints of the edges of ℓ. If also the label positions are fixed, the reference point of λ is limited to the ports of ℓ.

We call the geometric composition $\gamma = (\lambda, \ell, \pi(\ell))$ of a leader λ and a placed label ℓ attached to each other at the port $\pi(\ell)$ a *callout*. A set $\mathcal{L}$ of callouts is called an *external labeling* of the feature set F; for brevity we also call it *labeling*. We say that $\mathcal{L}$ is *plane* or *crossing-free* if no two callouts in $\mathcal{L}$ intersect. Typically the goal is to find a labeling that contains a callout for each feature. However, if the space for the placement of the labels is strongly restricted, finding a labeling for a subset of F can be a possible solution.

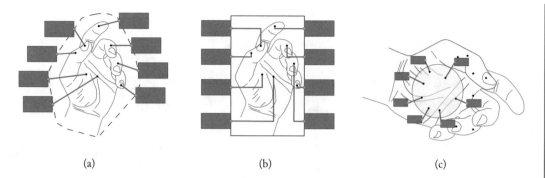

(a)   (b)   (c)

Figure 2.4: Admissible positions of labels. (a) Contour labeling: positions of reference points are restricted to a predefined contour (dashed). (b) Boundary labeling: positions of reference points are restricted to a rectangle. (c) Excentric labeling: the labeled features are contained in a circle (blue), while the labels are placed around the circle.

2.2   DISTINCTIVE FEATURES

In this section we discuss three distinctive features of external labeling that are used to characterize the different labeling models found in the external labeling literature: the position of labels, the type of the leaders, and the setting of the labeling, namely static or dynamic. These are not the only distinctive features of external labeling. For instance, the choice of colors and the content and orientation of the labels are equally important to define the visual result. However, from the algorithmic perspective of this book their relevance is not so central. Instead, we cover such aspects in Chapter 3.

2.2.1   ADMISSIBLE POSITIONS OF LABELS

The admissible positions of the reference points of the leaders and accordingly of the labels significantly influence the overall appearance of the labeled figure. While in general models for external labeling the reference points may be placed anywhere in the labeling area, which we call *free external labeling*, we distinguish the following more restrictive variants.

Contour labeling   In the model of *contour labeling* the positions of the reference points are restricted to a predefined contour C_A; see Figure 2.4(a). This contour is often chosen in such a way that it roughly matches the shape of the illustration in the background layer. This is a common technique to achieve a label placement that blends in with the illustration.

Boundary labeling   An even more restrictive model is that of *boundary labeling*, which requires that the image region R is a rectangle and C_A coincides with the boundary of R; see Figure 2.4(b). Hence, excluding the corners of R for the placement of reference points, each label intersects exactly one edge of the image boundary in more than one point; we say that the label *touches*

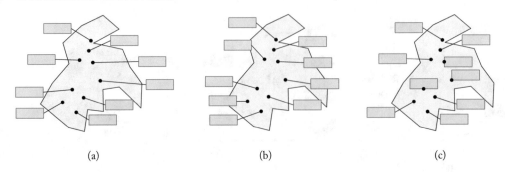

(a)   (b)   (c)

Figure 2.5: The strictness of external labelings. (a) In a *strict* external labeling all labels are strictly placed in the exterior of the image region. (b) In a *non-strict* external labeling some labels may overlap with the image region. (c) In a *mixed* labeling some labels are placed externally, while others are placed internally.

that edge. Depending on the number of edges that are touched by labels, we distinguish *1-sided*, *2-sided*, *3-sided*, and *4-sided* boundary labeling models. We observe that for 2-sided boundary labeling models the edges that are touched by labels are either *adjacent* or *opposite* edges of R; Figure 2.4(b) shows a 2-sided boundary labeling with opposite edges touched by labels.

Excentric labeling   In *excentric labeling*, only features in a restricted region of the illustration are labeled and the labels are placed outside that region; see Figure 2.4(c). In spite of labels possibly overlapping the actual illustration in the background, we consider excentric labeling to be a special case of external labeling, because the labels are still clearly separated from their features. In excentric labeling the restricted region is typically described by a circle of fixed radius and implements the metaphor of a *lens* that can be moved over the illustration to explore its details. Alternatively, one can think of this restricted region as a *focus region* of the user, i.e., the user is interested in the information contained in a specific region of the underlying background image. In the literature, two main variants are found: labels that are placed freely in the surroundings of the focus region and the special case of contour labeling requiring that all labels are placed along the boundary of the focus region. Moreover, in case that the leaders are *radial*, i.e., they all are part of a spokes emanating from the center of the focus region, the labeling is called a *radial labeling*.

Generally speaking, boundary labeling is commonly applied in labeling static background images (e.g., geographic maps), whose boundary is usually rectangular. Contour labeling, on the other hand, finds applications in scenarios, in which the silhouette of the underlying background image is more complex (e.g., technical drawings or anatomical drawings). Finally, the excentric labeling paradigm becomes useful in user-driven labelings, where the requirement is to convey the information that the user requests.

Some approaches relax the requirement that the labels must not overlap the image region. We distinguish *strict* and *non-strict* external labeling; see Figures 2.5(a) and 2.5(b), respectively. In the non-strict case the labels may partly overlap the image region in order to optimize other placement criteria. In case that some labels are placed internally such that they touch their corresponding features, we call the labeling a *mixed labeling*; see Figure 2.5(c). If not stated otherwise, we consider strict external labeling throughout this book.

2.2.2   LEADER TYPE

The model that is introduced in Section 2.1 allows general curves for leaders without any restriction to their shape. However, in actual externally labeled figures only leaders with simple, schematic shapes, such as straight lines and polylines, are used in order to avoid cluttered illustrations and to sustain legibility. In rare cases, *curved* leaders such as Bézier arcs are used. Next, we present a systematic classification of different polyline leader types. Let λ be a leader that consists of k segments, which are ordered from its site to its reference point. We describe the shape of λ by a string $z \in \{s, r, d, o, p\}^k$ of k symbols, where the i-th symbol of z describes the orientation of the i-th segment of λ. In the following, we list the meaning of each symbol; see Figure 2.6 for an illustration.

s   An s-segment is a straight-line segment with unrestricted slope.

r   An r-segment lies on a ray that emanates from a given center point M.

o   An o-segment is orthogonal to a reference line or line segment m.

p   A p-segment is parallel to a reference line or line segment m.

d   A d-segment l (for a given angle $0° < \alpha < 90°$) is a diagonal segment defined relative to its preceding segment l^- and its succeeding segment l^+ as follows. The turning angle between l^- and l as well as the turning angle between l and l^+ is $\pm\alpha$, where the turning angle of two segments sharing a common endpoint is the minimum angle in which one has to rotate one of the two segments around their common endpoint to reach the second.

Due to their shape, leaders consisting only of o- and p-segments are called *orthogonal* leaders, when the reference line or line segment is either vertical or horizontal. If one additional type of d-segments is allowed, then the resulting leaders are called *orthodiagonal*. In the special case, in which the turning angles of each d-segment with its preceding and succeeding segments are $\pm 45°$, the obtained leaders are called *octilinear* following the established terminology in network visualization.

In models for boundary labeling, the shape of a leader is typically described with respect to the edge of the boundary rectangle R that is touched by the leader's label. Following this convention, Figure 2.7(a) illustrates the most frequent leader types in the literature. The labels

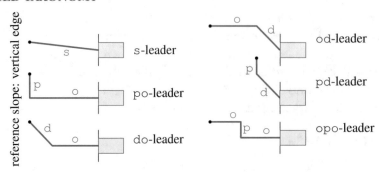

Figure 2.6: Notation used for describing different types of leaders.

A, E, F, G, and H are connected to s-leaders. More precisely, while A is connected to an arbitrary s-leader, the labels E, F, G, and H have r-leaders, because their straight-line segments lie on rays emanating from the center M. Further, the labels D and M have o-leaders.

The remaining labels B, C, I, J, K, and L have leaders with more than one segment. Assuming again that the reference line segment of each leader is the side of the rectangle containing its label, B has a po-leader, C has an opo-leader, and L has an po-leader; recall that the segments of a leader are ordered from its site σ to its reference point a. Further, label I has a do-leader, label J has a pd-leader, and label K has a do-leader. In all three cases the d-segment has a turning angle α with respect to its preceding and succeeding segment, respectively; with rare exceptions $\alpha = 45°$ is used for d-segments in the literature.

In some applications multiple features can share the same label, i.e., a single label is connected to multiple features via multiple leaders; see Figure 2.7(b). Using this labeling style the reader can easily identify features with the same meaning. Further, it saves space in the labeling area. We refer to this labeling technique as *many-to-one labeling*. In addition, by connecting all leaders to the same reference point of the label, the leaders can be bundled; we call the set Λ of these leaders a *hyperleader* of the label; see Figure 2.7(c). The literature on labeling with hyperleaders requires that no two leaders in Λ intersect apart from a common suffix. This leads to the desired visual impression that Λ is a single leader forking to multiple sites.

2.2.3   STATIC OR DYNAMIC LABELING

Interactive visualizations that change over time frequently use external labeling approaches. For example, Haunert and Hermes [55] consider the application of exploring a digital map by panning a lens over the map. For the point features within the lens they create an excentric labeling with the labels lying on the boundary of the lens; see Figure 2.8(a). As another example, consider a visualization system that provides the interactive exploration of a 3D model and that uses external labeling to explain the features of that model; see Figure 2.8(b). When the user changes the view of the 3D model, the labeling changes correspondingly. From a technical point of view,

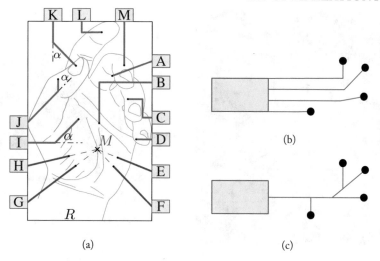

(a)

(b)

(c)

Figure 2.7: Illustration of different types of leaders: (a) Leader types mostly used in literature. (b) A label connected to multiple sites. (c) A hyperleader, i.e., a set of leaders having the same reference point, but different sites.

this process can be seen as an animation consisting of a temporal sequence of 2D images; see Figure 2.9. In case of a 3D model, this sequence is created by projecting the 3D model onto the screen space. For each of these 2D images, an external labeling is created to name the model's features. Hence, we obtain a temporal sequence of external labelings; we call this sequence a *dynamic labeling*. In contrast, we call a single labeling in that sequence a *static labeling*.

Hence, we can distinguish models for *static external labeling* that aim at individual labelings, and models for *dynamic external labeling* that aim at sequences of labelings. Models for dynamic external labeling may define constraints enforcing temporal coherence between consecutive labelings in order to avoid distracting effects such as flickering and jumping labels. For example, the red labels in Figure 2.9 *flicker* when going through the sequence from left to right. In particular, the red label with the black square and the red label with the white circle are displayed alternatingly. In contrast, the blue labels do not flicker.

Finally, we note that in most of the literature on labeling of 3D models the labels are placed in the 2D projection space, which is also our default here, but there are also some papers that place labels in the object space, e.g., [27, 89].

2.3   OPTIMIZATION PROBLEM

Next, we introduce the formal definition of the external labeling problem. As input we are given a drawing area $D = (R, A)$ partitioned into the image region R and the labeling region A, a feature set F, and a set M of model parameters to be specified below. We call the tuple $I = (D, F, M)$

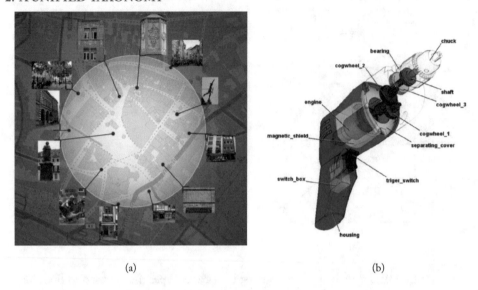

(a)   (b)

Figure 2.8: Examples of dynamic labelings. (a) A lens for interactively exploring point features on a digital map. (b) A 3D model that can be explored by interactively changing the perspective.

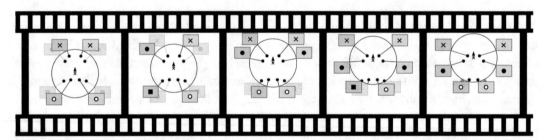

Figure 2.9: Temporal sequence of static labelings. A lens (black circle) is moved over a set of point features creating a sequence of frames. For each frame a static radial labeling is created; grey labels are not displayed to the user. Going through the sequence from left to right, the red labels flicker, while the blue labels are temporally coherent.

an *instance* of external labeling. A labeling $\mathcal{L}$ of F that satisfies all model parameters of M is called a *feasible* labeling of I. We denote the set of all feasible labelings of I by $\mathcal{S}$.

The model parameters are a set of layout-specific criteria that must be satisfied. Typical parameters are the selected leader types, site, port, and reference point constraints, and hard constraints that restrict the mutual interplay of the callouts or their interference with the illustration in the background layer. While the leader types have been specified in Section 2.2, the site, port, and reference point constraints define the admissible placement of the site (fixed, sliding, free), port (fixed, sliding), and reference point (fixed, sliding, free). Examples of hard constraints are crossing-free leaders, non-overlapping labels, and avoidance of obstacles in the illustration. The actual set of model parameters strongly depends on the application and is determined by the designer; see Chapter 3 for some commonly used guidelines.

Most commonly, external labeling is seen as an optimization problem to find the best among all feasible labelings (or at least a reasonably good one). Hence, it remains to define a cost function $c : S \to \mathbb{R}^+$ that measures the quality of the feasible labelings in S.

External Labeling
Input: An instance $I = (D, F, M)$ and a cost function $c : S \to \mathbb{R}^+$.
Output: An optimal labeling $\mathcal{L} \in S$, i.e., $c(\mathcal{L}) \leq c(\mathcal{L}')$ for each $\mathcal{L}' \in S$.

A concrete example of this problem is 1-sided boundary labeling with opo-leaders, uniform labels, and fixed ports minimizing the total number of bends. In particular, in that case we require that the image region R is a rectangle and that the ports of the labels lie on the boundary of R. Similarly, other variants such as contour and excentric labeling can be modeled. In the following, we list few of the most important optimization criteria; see also Section 3.2 for a detailed discussion.

- **Leader crossing minimization.** The most natural requirement in labelings with leaders is the avoidance of crossings between leaders, since crossings result in visual clutter. However, in scenarios with a large number of features to be labeled it is not always possible to guarantee the existence of a labeling without crossings between leaders. In such scenarios, one is naturally seeking for labelings with the minimum number of crossings to reduce visual clutter.

- **Leader length minimization.** Minimizing the length of the leaders is considered to be an important optimization criterion which facilitates unambiguity. A labeling with small leader length guarantees short distances between sites and labels, and therefore increases readability. It also minimizes the ink used to plot the leaders, and thus also the amount of overplotting of the image region.

- **Leader bend minimization.** If a leader contains several bends, then it is usually difficult for the human eye to follow it (especially when the bends define sharp changes of direction). Also, labelings with many leader-bends are usually cluttered. Hence, both the number of bends per leader and the total number of leaber bends of a labeling should be kept small.

- **Label number maximization.** In case that the labeling region is not large enough to host the labels of all features, maximizing the number of placed labels promises crossing-free labelings with high information density. In the simplest variant each label is rated equally, while more sophisticated objectives also take the importance of the features into account.

Depending on the concrete choice of the instance I and the cost function c, the external labeling problem can be solved in polynomial time for restricted cases but easily becomes NP-hard for general settings [10, 57, 70]. Two research directions have evolved. The first direction considers the external labeling problem from an *algorithmic perspective* focusing on the development of algorithms that solve restricted variants such as boundary labeling in polynomial time. The second direction considers the external labeling problem from an *applied perspective* focusing on the development of fast heuristics that can be deployed in visualization systems. In recent years there are efforts toward joining both directions.

CHAPTER 3

Visual Aspects of External Labeling

In this chapter we discuss various visual aspects that affect the aesthetic quality and usability of external labeling. In general, it is far from obvious which type of external labeling is the best choice and this highly depends on the application. We discuss several important criteria that must be taken into account when deciding on a type of external labeling. We distinguish between the *style* of the labeling and the actual *placement* of the labels based on it.

3.1 STYLE

The visual appearance of a callout in an externally labeled illustration depends on many aspects that influence the legibility and readability, as well as the overall aesthetics of the illustration to various degrees. In this section, we give a high-level overview of such aspects (that are not related to algorithmic aspects of optimizing the positioning of labels and leaders). Figure 3.1 illustrates the visual effects of these design choices.

The choice of the leader type (e.g., straight-line, polyline, or curved; see Chapters 5 and 6) has a strong influence on the overall appearance; it also has an influence on the available techniques to produce the actual labeling, as we discuss in Chapter 4. In terms of readability, it is important to consider the overlay of the illustration in the background layer and the leader curves in the labeling layer. One must ensure that lines in the background layer and leader lines can be easily differentiated [86], which means that in illustrations with arbitrary and irregular shapes (e.g., maps or anatomy drawings such as Figures 1.2 and 3.1), straight and polyline leaders are preferable, whereas in illustrations with many straight lines of regular slopes (e.g., floorplans), curved leaders, or straight and polyline leaders with non-aligned slopes should be considered [103].

In addition, multiple applications have their default leader type. For labelings in atlases of human anatomy or in dynamic settings (e.g., augmented reality) straight-line leaders are commonly used. On the other hand, do-leaders and po-leaders are frequently used in info graphics, technical drawings, and maps if they match the style. Finally, there are some special applications such as text annotation where opo-leaders naturally fit.

Richards [86] presented a list of best-practice guidelines originally for labeling black-and-white technical drawings in aerospace engineering, but most of them should also generalize to other domains.

The original version of this chapter has been revised. The Figure 3.3 has incorrectly updated in the chapter 3 of this book and that has now been corrected. The correction to this chapter is available at https://doi.org/10.1007/978-3-031-02609-6_8

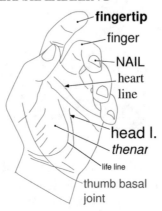

Figure 3.1: Different examples of rendering and typesetting callouts.

- Leader lines should be straight wherever possible and should consist of a black line and right next to it, on one side, a thin white line to separate the leader from the background.

- Leader lines (for area features) should terminate with a terminator symbol—either a small dot if the feature is large enough, or an arrowhead for small features.

- Polyline leaders producing staircase effects, possibly for aesthetic reasons, are acceptable as long as there are not too many leaders, but should be avoided for contexts where safety or efficiency of use are relevant.

The graphical rendering of leader lines themselves is also of relevance for the visual quality of a labeled image. Wu and Dalal [104] considered different quality feature of rendering lines on electronic displays, including the raggedness, blur, waviness, and lumpiness of lines, depending on the method of digitizing Euclidean straight lines and using anti-aliasing effects.

In addition to the leader, the label itself needs to be properly configured and displayed, e.g., in places with low salience [52]. While horizontally aligned labels are most common and easiest to read, slanted labels (e.g., diagonally or vertically aligned) consume less space when placing them above the figure. The question of showing a label as just the text itself, as the text on some colored background region or with an explicit bounding box, strongly influences the visual style as well. In situations, where the text is well readable without background or bounding box, this is usually sufficient and preferred by designers. In some dynamic applications, such as augmented reality, the background behind each label is unknown and hence lightly colored background boxes with or without their frame are commonly used. Note that the color of the background box or the frame can additionally be used to indicate semantically related groups of labels.

Another important consideration is the actual typesetting of the label text, which must be done appropriately. For example, Degani [32] gave a detailed list of guidelines and principles for typesetting of flight-deck documentation, whose use of text is comparable to labeling. We mention here the most important ones.

- While one can choose from a large variety of fonts, which should match the context of the illustration, for labels and short annotations generally sans-serif fonts are more legible than serif fonts.

- The selected font size depends on the estimated viewing distance and can be calculated according to standard guidelines. The same is true for non-textual labels such as icons.

- Lowercase words are more legible than uppercase words.

- Regular face is more legible than italic or bold face, which should only be used for highlighting specific words.

- Finally, a black font over white or lightly colored background has the best legibility.

For the text itself, one consideration is whether abbreviations, line breaks, or even omission of the text should be applied in order to make the text fit. In principle, a shorter label as a substitute when needed (e.g., acronym) might be helpful, as suggested by Shimabukuro and Collins [87] who presented a method for on-demand abbreviations of text labels.

3.2   PLACEMENT

From a visual point of view the general labeling style surely influences the appearance of the image most. However, once the labeling style has been fixed, the actual placement of the labels comes to the fore. Choosing the most legible and visually pleasing one among all feasible placements is a complex task that requires elaborate algorithmic techniques when done automatically, and a good balance of possibly contradicting placement criteria.

Static labelings   For static labelings, we have identified the following decisive criteria based on literature (e.g., [3, 77, 80]) that justifies them by interviews with experts and manual analysis of labeled drawings [80]. Figures 3.2 and 3.3 illustrate the effects of adhering to the criteria (left) and violating them (right).

C1 *The leaders have short length* to ease the association between features and labels. Shorter leaders increase the proximity, reduce the need for eye movement, and are in line with Tufte's principle of minimizing non-data ink [91].

C2 *The number of label overlaps is small* to reduce visual clutter and ambiguities and improve readability.

C3 *The number of leader crossings is small* to reduce visual clutter and ambiguities, following general principles in graph layout [85].

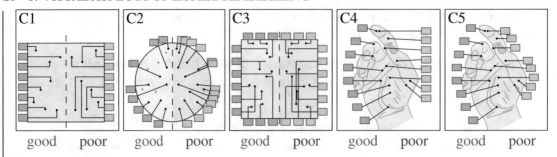

Figure 3.2: Illustration of labeling criteria CC1–CC5.

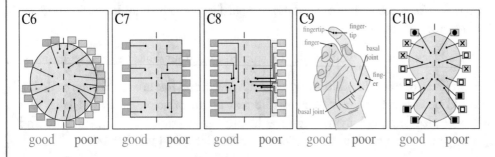

Figure 3.3: Illustration of labeling criteria CC6–CC10.

C4 *The labels mimic the shape of the image* to guarantee a clear distinction between labeling layer and background layer and to obtain labelings that blend in with the image.

C5 *The labels are distributed evenly* to avoid dense regions within the labeling layer and ensure a homogeneous appearance of the labeling.

C6 *The directions of the leader segments comply with a set of preferred directions* to avoid complicated shapes, support uniformity, and create visual contrast to unrestricted curves in the background image.

C7 *The leaders have a small number of bends* to sustain legibility and support readability by easing the association between features and labels, again following general principles in graph layout [85].

C8 *There is sufficient space between leaders* to reduce ambiguities between leaders.

C9 *Labels consist of single text lines if possible* to increase legibility.

C10 *Labels that are semantically related are grouped* to support comprehensibility.

We stress that there is not a clear, empirically confirmed ranking among these criteria, which is one of the research challenges discussed in Chapter 7; their respective relevance and relative priorities also depend on the application domain or even on the image to be annotated. Still they are supported by general labeling and drawing principles such as proximity, planarity, straightness/low detour, and small ink-ratio [40, 58, 85, 91, 97]. We also stress that some of these criteria might be in conflict with each other. As a result, a good labeling is usually a compromise of different criteria.

Dynamic labelings   Dynamic labelings are commonly found in data visualization, in 3D immersive environments (e.g., augmented reality and virtual reality) as well as in view management systems, in which the user can interactively explore 3D models. Here, the background layer is no more static but dynamic, as it may change over time. Hence, very often, special considerations are needed in order to distinguish the labels and the leaders from the background [43, 52, 66], to appropriately arrange the labels without obscuring the background layer [52] or even to filter the information being displayed so to achieve such placements [59]. On the other hand, the actual placement of the labels becomes an *online* problem for which the labeling algorithm needs to create a dynamic labeling without exactly knowing the upcoming changes.

As the frames of a dynamic labeling form a sequence of static labelings, the criteria for the static case also apply for the dynamic case, but are extended by the general requirement of temporal coherence between consecutive frames; see, e.g., [75]. Following general dynamic map labeling principles [8], *flickering* and *jumping* labels should be prevented, while optimizing and preserving the placement criteria listed above. The proposed approaches found in literature can be categorized into two groups [75].

The approaches of the first group consider continuously moving labels, which clearly prevents flickering and jumping labels. However, the placement criteria such as overlap-free labelings and sufficient spacing between labels, cannot be completely maintained. In contrast, the approaches of the second group use *hysteresis*, i.e., the displacement of the labels is delayed to stabilize the motion of the labels and to enforce selected placement criteria. However, this easily leads to labels that jump from one position to the next. Preserving the order of the labels and animations visualizing the changes are possible tools to soften this effect.

In general, animations have been considered as an important tool in interactive visualizations to visualize the transition from one state to the other [36]. We further note that Madsen et al. [75] showed in a detailed user study on labelings in augmented reality that the participants performed better in the conducted tasks when the labels were directly placed in the 3D space of the object instead of the 2D space. This result is further supported by the user study by Peterson et al. [83] showing that incorporating depth information into the labeling increased the performance of the participants in selection tasks. They particularly showed that this technique softens the negative effects of overlapping labels. Still, only few approaches [84, 89] that aim at a clear spatial distinction between objects and placed labels have been developed for external labeling in object space so far.

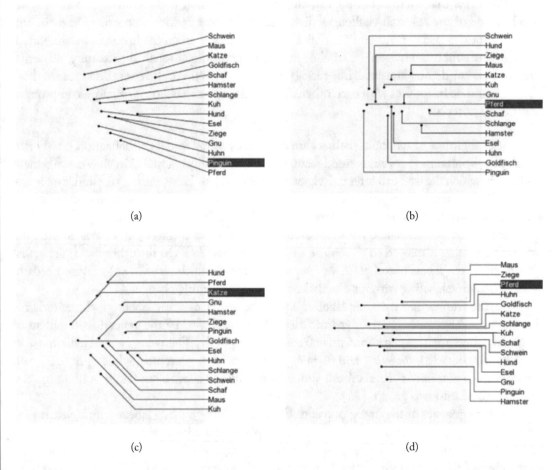

Figure 3.4: Stimuli used in the user study by Barth et al. [6] for assessing the readbility of different leader types. For each stimulus the participants were asked to click on the feature that is connected to the marked labeled. In order to avoid side effects no background image was shown. (a) s-leader, (b) po-leader, (c) do-leader, and (d) opo-leader.

3.3 EMPIRICAL STUDIES

In this section we briefly discuss the state of the art on different user and expert studies for empirically examining design decisions and evaluating labeling techniques.

The analysis of professional (handmade) drawings, e.g., technical drawings, visual dictionaries, medical illustrations, is a task that is neither complicated nor expensive and helps in extracting drawing criteria for a planned labeling technique. Niedermann et al. [80] extracted drawing criteria in a semi-automatic way from atlases of human anatomy to generate similarly looking labelings. Vollick et al. [94] proposed a system that automatically learns the drawing style from existing systems. While both approaches require that drawings in the specific style already exist, the analysis of existing drawings can also be used to justify default drawing criteria satisfied in almost all labeling styles [3, 53, 54, 77]. Moreover, interviews with domain experts help to justify the extracted criteria and to make further design decisions [28, 77, 78, 80].

Certainly the most extensive evaluation technique are user studies. In order to obtain useful results, an elaborated design, equipment and sufficiently many participants with different backgrounds are required. In research on external labeling, user studies are conducted to assess labeling approaches. Fekete et al. [37] investigated the question whether excentric labeling is a reasonable alternative for zooming. Čmolík and Bittner [28] compared their algorithm against label layouts created by humans. Madsen et al. [75] presented a study comparing different labeling approaches in dynamic settings; to the best of our knowledge this is the only investigation that takes multiple existing labeling algorithms into account. In contrast, Bertini et al. [19], Mühler and Preim [78], Pick et al. [84], Wu et al. [103], and Balata et al. [5] did not compare their results with other approaches, but investigated how participants perform when using their approaches and asked them concerning the experiences made when using their systems.

Instead of evaluating a concrete labeling approach, Barth et al. [6] conducted a user study to investigate general properties of external labeling, namely in this particular case the readability of different leader types; see Figure 3.4 for some example stimuli presented to the participants of the user study. Depending on the study, the participants were asked to conduct different types of tasks. As the main purpose of a labeling is to unambiguously relate features and labels to each other, the most often proposed task is to associate labels with their features and vise versa. This is either done by reading tasks [19] or by selecting particular labels [5, 6, 28]. Alternatively, the participants are asked to check the existence of particular labels [37] or to solve more complex tasks on the illustrations that require the use of the labeled features [105]. Only the user study by Wu et al. [103] used eye-tracking. Typical measures are the response times and error rates of the participants. Finally, except for the user study by Čmolík and Bittner [28], all presented studies take the participants' preferences and personal experiences with the labeling layouts into account. In particular, Pick et al. [84] assessed their approach by conducting an expert walkthrough in which the participants answered few questions after using their system.

CHAPTER 4

Labeling Techniques

In this chapter we discuss algorithmic techniques that are frequently applied in external labeling. We split them into non-exact (Section 4.1) and exact algorithms (Section 4.2). While exact algorithms solve instances of external labeling problems and provide an optimality guarantee for the solution, non-exact approaches refrain from such guarantees in favor of faster running time. In Section 4.3 we discuss complexity results mainly focusing on problems in external labeling that are NP-hard. We conclude this chapter in Section 4.4 with guidelines regarding the applicability of each technique.

4.1 NON-EXACT ALGORITHMS

First, we describe algorithmic techniques to solve instances of external labeling problems without necessarily yielding optimal labelings with respect to a cost function.

4.1.1 GREEDY ALGORITHMS

Greedy algorithms are simple local optimization strategies that iteratively construct solutions by *greedily* extend them with elements that promise the best improvement of the solution in each step. For an introduction to greedy algorithms see any algorithms textbook, e.g., [30, Chapter 17].

Characteristics

- Support general multi-criteria cost functions, but typically find only local optima.

- Hard constraints can be easily enforced, but may lead to unlabeled features or infeasible solutions.

- Easy to implement and typically fast in practice.

Sketch　　Greedy algorithms in external labeling can be summarized by the following scheme. For each feature a candidate set of callouts is parameterized; let $\mathcal{C}$ denote the union of all those sets. Starting with an empty labeling $\mathcal{L}$, the callouts are *greedily* added to $\mathcal{L}$ with respect to some given cost function and hard constraints. More precisely, the callout $c \in \mathcal{C}$ with lowest cost among all callouts in $\mathcal{C}$ is added to $\mathcal{L}$. Afterward, the callout c and all callouts that cannot be added to the current labeling $\mathcal{L}$ without violating a given hard constraint are removed from $\mathcal{C}$. The procedure is repeated until the candidate set $\mathcal{C}$ is empty. As a result of the procedure,

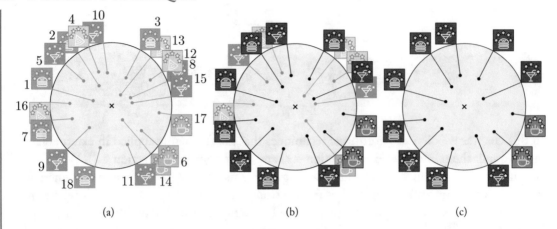

Figure 4.1: Illustration of a greedy algorithm applied on an instance for radial labeling. (a) All possible label candidates that can be chosen by the greedy algorithm. (b) The greedy algorithm successively selects candidates starting with high ratings. (c) The resulting overlapping-free labeling.

the labeling $\mathcal{L}$ is returned. Since in each step it is guaranteed that $\mathcal{L}$ is a labeling of the given instance satisfying all hard constraints, the resulting set $\mathcal{L}$ is also a valid labeling. However, it is not guaranteed that it is the cost-optimal labeling among all possible labelings. Further, depending on the hard constraints, in some cases it may happen that not all features are labeled, even if a complete labeling exists. This is because all candidate callouts of some feature may have already been removed from $\mathcal{C}$ before considering them in the greedy order.

Example   Figure 4.1 shows the application of a greedy algorithm creating a radial labeling. The point features lie within a circular area. As an use-case, suppose a digital map whose content can be explored by moving a lens over a map. Facilities (e.g., restaurants or cafés) that lie within the region of the lens are labeled with an icon and a one- to five-star rating.

In the first phase, the greedy algorithm creates for each contained point feature a callout whose straight-line leader is part of the spoke that emanates from the center of the lens and passes through the point feature. Further, it sorts the callouts by their star rating as primary key. Ties are broken by using the lengths of the leaders as secondary key: features that lie closer to the center of the lens are preferred. In the second phase, the greedy algorithm successively adds callouts to the solution based on the pre-computed ordering. In each step it excludes all callouts from the candidate set that overlap with the current selection. The result is an overlapping-free radial labeling. The example illustrates that the greedy algorithm does not necessarily yield the globally optimal solution. Assuming that the optimization goal is to maximize the total star rating of the selected labels, the solution can be improved by selecting the label candidates 4 and 5 instead of the label candidate 2.

4.1.2   FORCE-BASED ALGORITHMS

Force-based approaches are local optimization strategies that have been originally introduced in the field of graph drawing to construct layouts of graphs [34, 41]. According to this approach, a graph is modeled as a physical system with forces acting on it, and a good layout is obtained by simulating this system until an equilibrium state is reached. In the context of external labeling, the sites, labels, leaders, and features become the *objects* in the physical system that interact with each other by forces. The approaches mainly differ in which objects are fixed and which can move freely, the definition of the forces, and how an equilibrium is found.

Characteristics

- Supports general multi-criteria cost functions, but typically finds only local optima.

- Needs an initial labeling as starting point.

- Intuitive method as it is based on a mechanic analogy.

Sketch   A common technique to search for an equilibrium of the system is to iteratively apply two steps. Firstly, compute for each object o in the system the sum f_o of the forces that act on it. Secondly, move each object o along the direction of f_o by an amount that is proportional to the magnitude of f_o and to a predefined constant Δt that may change over time. If after a number of iterations the objects do not move any more or a maximum number of iterations is reached, the system is assumed to be in an equilibrium state. We note that the initial configuration of the system, i.e., the one in the first iteration, strongly influences the quality of the overall result of the procedure. For example Ali et al. [3] use this iterative approach as a post-processing step to improve a layout created beforehand. They introduce forces that optimize the angles between leaders and the distance between labels and their sites. Similarly as done in simulated annealing, they start with a large Δt and decrease it with each iteration. Hence, in each iteration the possible displacement of the labels is reduced.

Another approach defines potential functions for the objects based on force fields. The positions of the objects are then chosen such that they minimize the total potential of the objects. Hartmann et al. [53] introduce for each label a static force field such that the attracting force between a label and its feature increases with increasing distance. Further, a label is repelled by other objects and the border of the drawing area. The local minima of the potential function are found by using particles that move within this force field. Each local minimum is interpreted as a label candidate of the same feature. Multiple labels are placed iteratively. In each iteration a label is placed and fixed at its preferred position in its force field as described beforehand. For the remaining labels their force fields are adapted incorporating the newly placed label as repulsive force.

Example   Figure 4.2 shows the labeling of the cross section of a human brain. A simple force-based approach first creates an initial crossing-free labeling; see Figure 4.2(a). This particularly

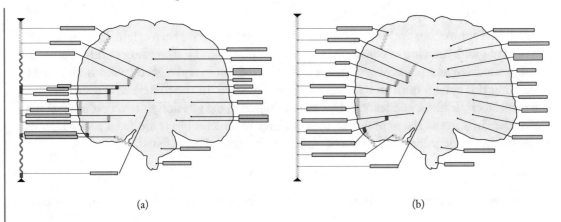

(a)   (b)

Figure 4.2: Illustration of two states of a force-based approach: (a) the initial crossing-free labeling and (b) the resulting labeling. The spring between a site and a label is colored green (if its length is close to its natural length), or yellow (if its length is relatively close to its natural length) or red (if its length is not close to its natural length).

requires to decide for each label on which side of the image it is placed. In the given example spring forces are introduced for evenly distributing the labels on each side of the image; see springs on the left hand side in Figure 4.2(a). For example, for two adjacent labels ℓ_1 and ℓ_2 one may introduce the spring-like force

$$f_{\mathrm{spring}}(\ell_1, \ell_2) = c_{\mathrm{spring}} \cdot \log \frac{d(\ell_1, \ell_2)}{l},$$

where l denotes the ideal spring length and where $d(\ell_1, \ell_2)$ denotes the distance between the centers of the labels ℓ_1 and ℓ_2. Further, repulsive forces between leaders and sites are introduced to avoid leaders that run closely by sites; see springs between sites and leaders in Figure 4.2(a). For example, for a leader λ and a closely located site p one may introduce the repulsive force

$$f_{\mathrm{rep}}(\lambda, p) = \frac{c_{\mathrm{rep}}}{d(\lambda, p)},$$

where $d(\lambda, p)$ denotes the distance between λ and p. Using an iterative approach the labels are moved along the contour minimizing the forces in total; see Figure 4.2(b).

4.1.3   MISCELLANEOUS TECHNIQUES

In this part we briefly mention meta heuristics and approximation algorithms as other well-known techniques that sporadically have been applied in the context of external labeling.

Meta heuristics are general techniques that serve as a template for solving optimization problems heuristically. In external labeling *simulated annealing* [68, 94] and *genetic algo-*

rithms [71, 102] have been considered. Both techniques have in common that they define neighborhoods of solutions by specifying (local) transformation rules to move from one solution to another. Based on these neighborhood structures, the search space is systematically explored, preferably moving toward better solutions, but sometimes also following worse intermediate solutions, in order to find an optimal solution, at least a locally optimal one.

Approximation algorithms give formal guarantees on the quality of the result. For example, Lin et al. [69] present approximation algorithms for minimizing the number of crossings between opo-leaders, i.e., they prove that the number of crossings in the constructed labeling is at most a constant times the number of crossings in a labeling with an optimal number of crossings. Bekos et al. [11] present 2-approximation algorithms for a mixed labeling problem maximizing the number of internal labels, i.e., they formally prove that the number of internal labels in the constructed labeling is at least half the number of internal labels in a labeling with the maximum number of internal labels.

4.2   EXACT ALGORITHMS

In this section we describe algorithmic techniques that are used to solve instances of external labeling problems exactly, i.e., with respect to a given cost function these techniques yield optimal labelings.

4.2.1   DYNAMIC PROGRAMMING

Dynamic programming is by far the most used technique in the context of external labeling (see also Tables 5.1 and 6.1). The general idea is to find an optimal labeling using recursion originating from the input instance. In each recursive step, the optimal labeling of the currently considered instance I is composed of optimal labelings obtained from smaller disjoint and compatible sub-instances of I. The base case of the recursion is reached when I does not contain any or just a single feature to be labeled and the optimal labeling of I is trivial. Storing the intermediate solutions in a table for subsequent look ups ensures that each sub-solution is computed only once, which keeps the running time polynomial. For an introduction to dynamic programming, see [30, Chapter 16].

Characteristics

- Allows general multi-criteria cost functions.

- Requires as a hard constraint that feasible labelings are crossing-free.

- Further hard constraints can be easily incorporated.

- Predominantly applied in boundary labeling (see Tables 5.1 and 6.1).

- Depending on the actual setting, dynamic programming tends to yield algorithms that have high asymptotic running time and storage consumption.

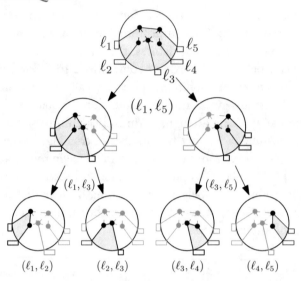

Figure 4.3: Decomposition tree of an instance for external labeling. The instance (blue) enclosed by the labels ℓ_1 and ℓ_5 (top) is decomposed into sub-instances enclosed by the labels ℓ_1 and ℓ_3, and ℓ_3 and ℓ_5, respectively (middle row). Each of them is again decomposed into two instances, which do not contain any point features anymore.

- Designing dynamic programming approaches requires problem specific insights.

Sketch   In the following, we sketch the technique in greater detail providing a general blueprint. Let $I = (D, F, M)$ be an instance of external labeling. We call an instance $I' = (D', F', M)$ a *sub-instance* of I if the drawing area D' is a sub-region of D and the features F' are a subset of the features in F. Further, two instances are *disjoint* if their drawing areas do not intersect.

The dynamic programming approach recursively constructs an optimal labeling $\mathcal{L}$ of I based on optimal labelings of disjoint sub-instances of I; see Figure 4.3 for an illustration. To that end, a set $\mathcal{I}$ of sub-instances of I is defined (including I) such that I can be recursively decomposed into instances from $\mathcal{I}$. More precisely, two instances, I' and I'', form a *decomposition* of I if

- I' and I'' are disjoint sub-instances of I,

- each feature of I is contained either in I' or in I'', and

- both I' and I'' contain fewer features than I.

The recursive decomposition of I is then described as a rooted, binary tree T over $\mathcal{I}$ such that

- each node of T is an instance of $\mathcal{I}$,

- each node of T has either no or two children,

- the children of a node form a decomposition of that node,

- I is the root of T, and

- the leaves of T are empty instances containing no features.

We call such a tree a *decomposition tree* of I. The actual choice of $\mathcal{I}$ strongly depends on the concrete problem setting and typically requires fundamental insights into the geometric structure of the problem. In particular, the set $\mathcal{I}$ is chosen such that there is a decomposition tree T with

$$
c_{\mathrm{OPT}}(I) = \begin{cases} c_I, & \text{if } I \text{ is a leaf,} \\ c_{\mathrm{OPT}}(I') + c_{\mathrm{OPT}}(I'') + c_{I',I''} & \text{if } I \text{ has children } I' \text{ and } I'', \end{cases}
$$

where $c_{\mathrm{OPT}}(I)$ denotes the cost of an optimal labeling of I, and c_I and $c_{I',I''}$ are the base costs of labeling I or merging I' and I'', respectively. The existence of such a decomposition tree directly leads to the following classic dynamic programming approach. We introduce a table X that contains for each instance $I \in \mathcal{I}$ an entry $X[I]$ representing the cost of an optimal labeling of I, i.e., $X[I] = c_{\mathrm{OPT}}(I)$. Since we consider a minimization problem, we can use the following recurrence to actually compute $X[I]$:

$$
X[I] = \begin{cases} c_I, & \text{if } I \text{ is empty} \\ \min_{(I',I'') \in \mathcal{D}_I} X[I'] + X[I''] + c_{I',I''} & \text{otherwise,} \end{cases}
$$

where $\mathcal{D}_I \subseteq \mathcal{I} \times \mathcal{I}$ is the set of all decompositions of I in $\mathcal{I} \times \mathcal{I}$. We observe that it suffices to consider decompositions from $\mathcal{I} \times \mathcal{I}$, because the decomposition tree only consists of instances from $\mathcal{I}$. Further, we compute each table entry only once—for example in increasing order with respect to the number of contained point features. Hence, each table entry $X[I]$ is computed in $\mathcal{O}(|\mathcal{I}|^2)$ time. Since there are $|\mathcal{I}|$ entries in X, we obtain $\mathcal{O}(|\mathcal{I}|^3)$ running time and $\mathcal{O}(|\mathcal{I}|)$ storage consumption for computing the costs of an optimal labeling of an input instance. Using a standard backtracking approach [30] that reconstructs the solution based on the decomposition tree, we also obtain an optimal labeling for that instance. It is crucial to keep $\mathcal{I}$ as small as possible, because the running time of the approach strongly depends on the size of $\mathcal{I}$. A common approach is to show that the drawing area of the instances can be specified by only few parameters over small domains and to use these parameters to appropriately define the subinstances.

Example   The majority of approaches that are found in literature specify the drawing area of an instance based on leaders (or on callouts if the positions of the labels are not fixed). For example, consider 1-sided boundary labeling for point features and do-leaders such that the

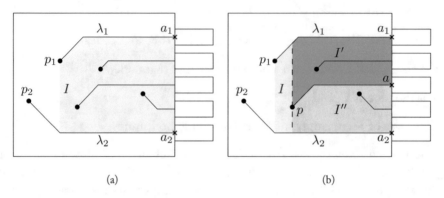

(a)   (b)

Figure 4.4: 1-sided instance of boundary labeling with do-leaders. (a) The instance I (blue drawing area) is specified by two point features p_1 and p_2, and the two reference points a_1 and a_2. (b) The leader λ with site p and reference point a decomposes the instance I into the two disjoint sub-instances I' (red drawing area) and I'' (orange drawing area).

label positions are fixed, the labels touch the right of the boundary rectangle R, and the ports of the labels are also fixed on the midpoint of their left edges; see Figure 4.4. Let A be the set of the induced reference points. We assume $|A| \geq |F|$, and that the cost function rates single callouts. Similar to Benkert et al. [18] we define the drawing area of a sub-instance of that problem setting by a simple polygon specified by two point features p_1, p_2 and two reference points a_1, a_2; see Figure 4.4(a). More precisely, the uniquely defined do-leader λ_1 connecting p_1 with a_1 bounds the area from above, and the do-leader λ_2 connecting p_2 with a_2 bounds the area from below. Further, the area is bounded from the right by the bounding rectangle and from the left by the vertical line that goes through the rightmost of the two sites p_1 and p_2. We assume that λ_1 and λ_2 neither belong to the drawing area nor to the instance and observe that all point features of such an instance lie to the right of the sites of both leaders. Such an instance is uniquely defined by the choice of the point features and reference points of λ_1 and λ_2. Thus, all such instances can be described by the tuples $(p_1, a_1, p_2, a_2) \in \mathcal{I} = F \times A \times F \times A$. Hence, $\mathcal{I}$ has size $\mathcal{O}(|F|^2|A|^2) = \mathcal{O}(|A|^4)$.

Benkert et al. [18] proved that for any input instance there is a decomposition tree over $\mathcal{I}$ as defined above. To see that, consider an optimal labeling $\mathcal{L}$ of an instance I in $\mathcal{I}$; see Figure 4.4(b). Let λ_1 and λ_2 be the leaders that bound I. We denote the point features and reference points of λ_1 and λ_2 by p_1, p_2 and a_1, a_2, respectively. If I does not contain any point features, the instance I is a leaf in the decomposition tree, and we have $c_{\text{OPT}}(I) = 0$. Otherwise, if I contains some point features, let λ be the leader with leftmost site in $\mathcal{L}$. We denote its site by p. The leader splits the instance into the two disjoint sub-instances $I' = (p_1, a_1, p, a)$ and $I'' = (p, a, p_2, a_2)$. The instances I' and I'' are the children of I in the decomposition tree. Further, we obtain $c_{\text{OPT}}(I) = c_{\text{OPT}}(I') + c_{\text{OPT}}(I'') + c(\lambda)$, where $c(\lambda)$ denotes the costs for

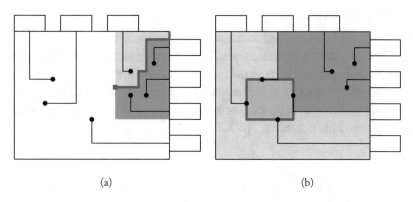

(a)   (b)

Figure 4.5: Illustration of separating two instances (orange and red). (a) Kindermann et al. [63] use an xy-montone curve to split a 2-sided instance into two 1-sided instances. (b) Bose et al. [22] use rectangles to separate instances from each other.

the leader λ, e.g., its length or number of bends. Hence, we can apply the dynamic programming approach as described above. The set $\mathcal{I}$ contains $n = \mathcal{O}(|A|^4)$ instances and each instance has $m = \mathcal{O}(|A|)$ decompositions; recall that we assume $|A| \geq |F|$. Hence, we obtain $\mathcal{O}(|A|^5)$ running time and $\mathcal{O}(|A|^4)$ storage consumption. For po-leaders, Benkert et al. [18] further show that an instance can be uniquely specified by two reference points improving the size of $\mathcal{I}$ to $n = \mathcal{O}(|A|^2)$, which yields a running time of $\mathcal{O}(|A|^3)$ and space consumption of $\mathcal{O}(|A|^2)$. Similar dynamic programming approaches have been considered for other leader-types; see Tables 5.1 and 6.1.

For the multi-sided case of boundary labeling the specification of the instances becomes more intricate. For labels on opposite sides Benkert et al. [18] specify instances for do-labelings using four articulation points and six point features obtaining $n = \mathcal{O}(|A|^{10})$ instances in $\mathcal{I}$. Further, they prove that it is sufficient to consider $m = \mathcal{O}(|A|^4)$ decompositions per instance. For po-leaders they improve the bounds for n to $\mathcal{O}(|A|^8)$ and for m to $\mathcal{O}(|A|^2)$.

While most approaches specify instances merely based on leaders, few approaches also use other separators. For example, Kindermann et al. [63] show for the 2-sided case with po-leaders and articulation points on adjacent sides that the instances can be described by an xy-monotone curve separating two 1-sided instances; see Figure 4.5(a) for an illustration. They show that it suffices to consider $n = \mathcal{O}(|A^2|)$ such instances and $m = \mathcal{O}(1)$ decompositions per instance to decide whether an instance admits a plane labeling in $\mathcal{O}(|A|^2)$ time and $\mathcal{O}(|A|)$ space. Requiring that the total leader length should be minimized, Bose et al. [22] show for the same problem setting that axis-aligned rectangles can be used to separate instances; see Figure 4.5(b) for an illustration. Using geometric properties of length-minimal labelings they show that their approach runs in $\mathcal{O}(|A|^3 \log |A|)$ time.

4.2.2   WEIGHTED MATCHING

Weighted matching has been applied successfully to instances of external labeling, where the objective is to minimize the total leader length, under few reasonable assumptions; see, e.g., [10, 13, 16]. The idea here is, given an instance of the external labeling problem, to construct an edge-weighted bipartite graph, such that a perfect matching of minimum cost for this graph translates back into a labeling of the original input instance, whose total leader length is minimum. For an introduction to matchings,, see, e.g., [2, Chapter 12].

Characteristics

- Mostly applies to boundary labeling problems where the objective is to minimize the total leader length [10, 13, 16].

- A post-processing step is usually required to make the computed labelings crossing-free.

- The running time of the algorithm is usually dominated by the time needed to solve the matching problem.

- Requires uniform labels placed at fixed positions.

Sketch   We sketch this technique for an instance $I = (D, F, M)$ of the 4-sided boundary labeling model, in which the feature set F consists of n point features, the leaders are of type-do, and both the positions of the labels within the labeling region A and the ports of the labels are fixed [16].

The main step in this technique is the definition of a weighted bipartite graph $G = (V_F \cup V_L, E, w)$, where $w: E \to \mathbb{R}^+$. The node set of G is defined as follows. For each point feature p of F, graph G contains a node $v(p)$ in V_F, and for each label candidate ℓ, graph G contains a node $v(\ell)$ in V_L. The edge set of G is then defined as the Cartesian product of V_F and V_L, that is, $E = V_F \times V_L$. In other words, graph G is complete bipartite. To complete the description of the definition of graph G, it remains to describe the weights of its edges. To this end, consider an edge e of G. By definition, edge e connects a node $v(p)$ in V_F to a node $v(\ell)$ in V_L. The weight $w(e)$ of edge e is defined as the length of the leader λ connecting point feature p with the port $\pi(\ell)$ of the label candidate ℓ. Equivalently, since our leaders are of type-s, the weight $w(e)$ of edge e is set to the Euclidean distance between point feature p and port $\pi(\ell)$.

A labeling $\mathcal{L}$ for instance I that minimizes the total length of the leaders corresponds to a minimum-cost perfect matching for G, which exists since G is complete bipartite; also, the triangle inequality ensures that no two leaders of $\mathcal{L}$ intersect, i.e., labeling $\mathcal{L}$ is plane. In general, the problem of finding a minimum-cost matching of a weighted bipartite graph can be solved in $\mathcal{O}(n^3)$ time using the Hungarian method; see, e.g., [65]. However, graph G has a special property that allows this problem to be solved more efficiently, namely, graph G is geometric (that is, drawn on the plane), and the weight of each edge corresponds to the Euclidean distance

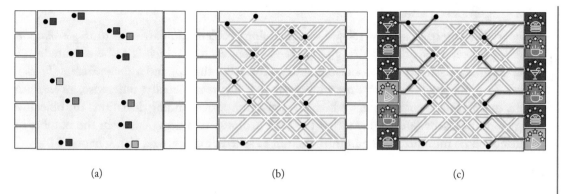

(a)   (b)   (c)

Figure 4.6: Illustration of the application of a matching algorithm: (a) the input labeling instance, (b) the construction of the complete bipartite graph between the point features and the labels, and (c) the labeling obtained by solving the minimum-cost perfect bipartite matching.

between its endpoints. For such graphs, the problem of finding a minimum-cost matching can be solved in time $\mathcal{O}(n^{2+\delta})$, where δ can be chosen arbitrarily small [1].

The technique can also be applied to labels with sliding ports with a minor modification, namely, that the weight of each edge $e = (v(p), v(\ell))$ of G must be the length of the *shortest* leader connecting point feature p with label ℓ. On the negative side, however, graph G is not geometric anymore, which implies that finding a minimum-cost matching for graph G now takes $\mathcal{O}(n^3)$ time.

We conclude this section by noting that this technique has also been applied to boundary labeling with opo-leaders [13, 16], and with a combination of od- and pd-leaders [10]. However, in these cases a post-processing step is required to eliminate potential crossings between leaders that may appear in the labeling $\mathcal{L}$ computed by the minimum-cost matching for graph G. It is also worth noting that for the case of opo-leaders and labels with fixed ports, the minimum-cost bipartite matching problem can be solved more efficiently as observed in [16]. This is due to the fact that the bipartite graph G is geometric with respect to the Manhattan distance, which allows for a solution in $\mathcal{O}(n^2 \log^3 n)$ time [92].

Example   Figure 4.6 shows an example of the 2-sided boundary labeling problem with uniform labels, fixed ports, and do-leaders. To minimize the total leader length a matching algorithm is applied. First a complete bipartite graph is created between the labels and the point features. An edge between a label ℓ and a point feature p corresponds to the uniquely defined do-leader connecting p with ℓ. The cost of the edge is defined as the length of that leader. Hence, a minimum-cost perfect matching corresponds to a boundary labeling with minimum total leader length.

4.2.3   SCHEDULING

Boundary labeling has also been related to scheduling problems, in particular to *single-machine scheduling*. Here, a set of n jobs $J_1, J_2, \ldots, J_n$ is given, each of which is to be executed on a single machine. Each job J_i is associated with a processing time ρ_i and a *time window* (b_i, d_i). If a job J_i is processed within its time window, then it incurs no penalty; otherwise, an *earliness* penalty or a *tardiness* penalty incurs, which is equal to the corresponding deviation. The objective is to determine a schedule, so that either the total earliness-tardiness penalty or the number of penalized jobs is minimized. For an introduction to scheduling see, e.g., [24, Chapter 1].

Characteristics

- Mostly applies to instances of 1-sided of boundary labeling, in which the objective is to minimize the total leader length or to minimize the total number of leader bends [14, 48].

- The technique has also been used to prove NP-hardness results [10, 57, 70].

- The main idea is that the height of each label corresponds to the processing time of a corresponding job, while the objective to be optimized is expressed as a sum of the earliness and of the tardiness penalties of the jobs.

Sketch   We sketch this technique on an instance $I = (D, F, M)$ of boundary labeling with type-opo leaders and sliding labels of non-uniform sizes that must be placed on one side on the boundary of the image region R, say the right side. For each $i = 1, 2, \ldots, n$, point feature p_i of I is associated with a job J_i, whose processing time is h_i and whose due time window is $(y_i - h_i, y_i + h_i)$, where y_i and h_i denote the y-coordinate of p_i and the height of the label of p_i, respectively.

Assume first that S^* is a schedule that minimizes the number of penalized jobs and consider the labeling $\mathcal{L}$ obtained by setting the y-coordinate of the lower-left corner of the label of the point feature p_i to the starting time of job J_i in S^*. Then, it is not difficult to see that $\mathcal{L}$ has the minimum number of leader bends, because for each $i = 1, 2, \ldots, n$ the leader connecting point feature p_i to its label is of type o (that is, without any bend) if and only if job J_i is completely processed in its time window in S^*. On the other hand, if S^* minimizes the total earliness-tartiness penalties, then a symmetric argument implies that labeling $\mathcal{L}$ is optimal in terms of the total length of the leaders. These simple linear-time reductions imply that finding 1-sided boundary labelings with sliding labels of non-uniform sizes that are either optimal in terms of the total number of leader bends or in terms of the total leader length takes $\mathcal{O}(n^2)$ and $\mathcal{O}(n \log n)$ time, respectively [14, 64].

We finally note that the above reduction has also been used to prove NP-hardness results mostly for non-uniform labels. The main idea here is that the order of the labels (and as a result the order of the jobs to be executed on the machine) may be variable when the leaders are not of type *opo*; see, e.g., [10, 57, 70]. Hence, finding a schedule that minimizes the total

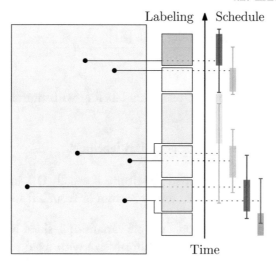

Figure 4.7: Reducing the 1-sided boundary labeling with opo-leaders to scheduling. The labels are associated with time windows illustrated to the right of the labeling. The obtained solution minimizes the total number of penalized jobs, and thus the total number of bends.

earliness-tardiness penalty or the number of penalized jobs becomes NP-hard, which gives rise to corresponding NP-hardness reductions.

Example Figure 4.7 shows an example for the 1-sided boundary labeling problem with sliding ports, non-uniform labels and opo-leaders. The scheduling algorithm is used to minimize the number of bends. Hence, for each point feature p_i an interval $(y_i - h_i, y_i + h_i)$ is introduced, where y_i and h_i denote the y-coordinate of p_i and the height of the label of p_i, respectively. The instance contains for each point feature p_i an interval I_i with length h_i such that its starting point coincides with the y-coordinate of the label of p_i. As argued above, minimizing the number of penalized jobs, i.e., jobs that are not entirely executed in their respective time window, also minimizes the number of bends.

4.2.4   PLANE SWEEP

Plane sweep is a well-known technique with several applications, mainly in computational geometry. This technique has been proven to be useful also in boundary labeling since several instances of this problem have purely geometric flavor (in particular, for uniform labels with fixed position and fixed ports). We stress, however, that the technique mostly applies when the goal is to find a crossing-free solution efficiently (that is, without taking into account any optimization criterion). A notable exception is the $\mathcal{O}(n \log n)$-time algorithm by Benkert et al. [18]

to compute 1-sided length-minimal boundary labelings with po-leaders. For an introduction to plane-sweep approaches see, e.g., De Berg et al. [31].

Characteristics

- Mostly applies to variants of boundary labeling with uniform labels, fixed reference points and fixed ports.

- Usually results in crossing-free boundary labelings.

- The running time of the obtained algorithms is usually $\Omega(n \log n)$ since a sorting of the input point features (e.g., from top to bottom or from left to right) is required.

Sketch   We describe this technique on a simple variant of 1-sided boundary labeling with s-leaders, in which the external labels are of uniform size with fixed reference points and ports aligned on a single side, say the right side, of the image region R. The algorithm is incremental, as it processes the labels from bottom to top in order to compute a labeling in which no two leaders cross [16].

More precisely, let $\pi_1, \pi_2, \ldots, \pi_n$ be the ports of the labels as they appear from bottom to top. Then, for $i = 1, 2, \ldots n$, the i-th port π_i is connected via an s-leader to the first un-labeled point feature that is hit by a ray r_i that emanates from π_i, initially points vertically downward, and is rotated around π_i in clockwise order. Suppose that a crossing occurs between any two leaders. But then the rotating ray would have found the corresponding point features in the reverse order, which contradicts the fact that the two leaders cross. A straightforward implementation yields a time complexity of $\mathcal{O}(n^2)$, which can be improved to $\mathcal{O}(n \log n)$ using standard techniques from computational geometry.

We conclude the description of this technique by mentioning that the algorithm presented above can be extended to compute corresponding 4-sided boundary labelings, under the same set of restrictions [16]. The idea is to partition the image region R into a particular number of convex non-overlapping sub-regions (using again appropriate rotating lines), such that the number of point features in the interior of each sub-region equals the number of labels on its boundary, which implies that the 1-sided algorithm can be applied to each of the computed sub-regions independently.

Example   Figure 4.8 shows an example for the 1-sided boundary labeling problem with uniform labels, fixed ports, and po-leaders. For minimizing the total leader length a sweep algorithm is applied. It works similarly as the approach sketched above. The horizontal sweep line moves from top to bottom. Each time it encounters a point feature, it adds the feature to a queue sorted by the x-coordinate of the point features in increasing order. Each time it encounters a label it removes the rightmost point feature from the queue and connects it with the label.

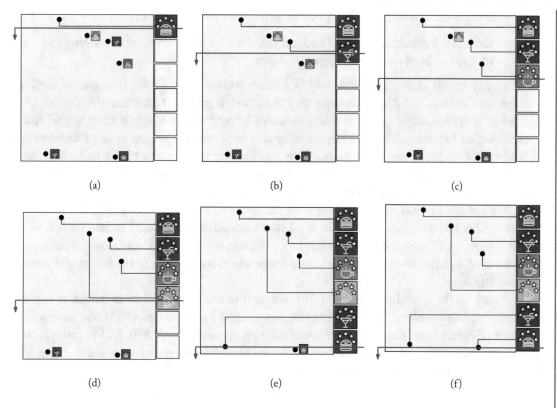

Figure 4.8: Illustration of a sweep algorithm to compute a 1-sided po-labeling with minimum leader length.

4.2.5   MATHEMATICAL PROGRAMMING

Mathematical programming is a general global optimization approach. A mathematical program consists of a set of variables, a set of inequalities, and an optimization function. The goal is to find an assignment of the variables such that all inequalities are satisfied and the value of the optimization function is (depending on the application) minimized or maximized. Often, these approaches make use of integer or binary variables. While this particularly provides the possibility of easily formulating combinatorial problems, integer variables, generally speaking, turn the optimization problem into an NP-hard one. Still, powerful solvers for mathematical programs have been developed, which can be used to solve a broad variety of instances in reasonable time. For an introduction to mathematical programming see, e.g., [93].

Characteristics

- General technique that can be used to formulate many (NP-hard) optimization problems.

- Solvers are available, but their running times are hard to predict.

- Common technique to obtain optimal solutions for the purpose of evaluating optimality gaps of heuristics for the same problem.

Sketch   We briefly sketch a basic variant of a mathematical program for the external labeling problem (see Section 2.3). More precisely, we present an integer linear programming formulation (ILP), i.e., a mathematical program that consists of integer variables, linear inequalities, and a linear objective function. Alternatively, one may also use quadratic programming formulations. Instead of linear cost functions they also allow multiplying two variables in order to formulate quadratic cost functions. While this supports modeling more complex problems, this usually also comes with a significantly higher running time in practice.

We assume that we are given a set F of features and for each feature $f \in F$ a set L_f of callout candidates; we denote the union of all those candidate sets by L. Further, let $c \colon L \to \mathbb{R}$ be a cost function that rates each candidate in L. We aim at a labeling $\mathcal{L} \subseteq L$ such that for each feature $f \in F$ exactly one callout $\gamma \in L_f$ is contained in $\mathcal{L}$ and $\sum_{\gamma \in \mathcal{L}} c(\gamma)$ is minimized among all those labelings.

Based on the work by Barth et al. [6], we use the following ILP formulation to express this optimization problem. We note that this is a standard formulation to express assignment problems. For each candidate $\gamma \in L$ we introduce a binary variable $x_\gamma \in \{0, 1\}$. We interpret the variable x_γ such that $x_\gamma = 1$ if the callout is selected for the labeling and $x_\gamma = 0$ otherwise. We further introduce the following linear constraints:

$$x_\gamma + x_{\gamma'} \leq 1 \text{ for each } \gamma, \gamma' \in L \text{ that exclude each other} \tag{4.1}$$

$$\sum_{\gamma \in L_f} x_\gamma = 1 \text{ for each feature } f \in F. \tag{4.2}$$

Constraint (4.1) ensures that if two callouts exclude each other due to some given hard constraint (e.g., because they intersect), at most one of them is added to the labeling. Further, Constraint (4.2) enforces that for each feature exactly one candidate is selected. Subject to Constraints (4.1) and (4.2), we minimize the total costs $\sum_{\gamma \in L} c(\gamma) \cdot x_\gamma$. The resulting labeling is then defined as $\mathcal{L} = \{\gamma \in L \mid x_\gamma = 1\}$. We note that Barth et al. [6] use a similar formulation to minimize the total leader length in boundary labeling.

Alternatively, if not all features can be labeled one may also maximize the number of labeled features or the total weight of all features. In that case, Constraint (4.1) is replaced by the following constraint:

$$\sum_{\gamma \in L_f} x_\gamma \leq 1 \text{ for each feature } f \in F. \tag{4.3}$$

Further, we maximize $\sum_{\gamma \in L} w(\gamma) \cdot x_\gamma$, where $w \colon L \to \mathbb{R}$ returns a weight for each candidate. Setting $w(\gamma) = 1$ for each candidate $\gamma \in L$ this corresponds to maximizing the number of selected labels.

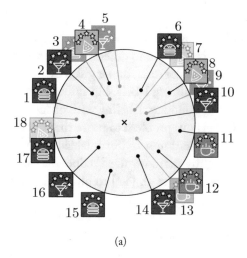

variables

$x_1, \ldots, x_{18} \in \{0, 1\}$

weights

☆☆☆☆☆ w_{15}

☆☆☆☆☆ w_{10}, w_{11}, w_{18}

☆☆☆☆☆ w_7

☆☆☆☆☆ $w_5, w_8, w_9, w_{14}, w_{16}, w_5$

☆☆☆☆☆ $w_1, w_2, w_3, w_4, w_6, w_{12}, w_{17}$

objective

$\max \sum_{i=1}^{18} w_i \cdot x_i$

constraints

$x_2 + x_3 \leq 1$	$x_7 + x_8 \leq 1$	$x_{12} + x_{13} \leq 1$
$x_3 + x_4 \leq 1$	$x_8 + x_9 \leq 1$	$x_{13} + x_{14} \leq 1$
$x_4 + x_5 \leq 1$	$x_9 + x_{10} \leq 1$	$x_{17} + x_{18} \leq 1$
$x_6 + x_7 \leq 1$		

(a)   (b)

Figure 4.9: Illustration of a mathematical programming formulation for a labeling instance (see also Figure 4.1), in which each label has a single label candidate and a weight (from 1–5). The formulation results in an overlap-free labeling (guaranteed by constraints) of maximum total weight (due to the objective function).

Example   Figure 4.9 shows an example of an instance for excentric labeling where the leaders are radial and the labels touch the boundary of a given circular focus region. Each feature i with $1 \leq i \leq 18$ has exactly one candidate γ_i. Further, candidate γ_i has weight w_i expressing the star rating of the feature. Maximizing $\sum_{i=1}^{18} w_i \cdot x_i$ subject to Constraint (4.1), we obtain a labeling that maximizes the total star rating. In the special case that each feature has exactly one candidate, we can omit Constraint (4.3).

4.3   COMPLEXITY RESULTS

In contrast to point labeling problems with internal labels, where the majority of the problems turn out to be NP-hard, several external labeling problems admit polynomial-time solutions, as we have seen in the previous section. In external labeling the problems become usually NP-hard when the labels are non-uniform or their positions are not fixed. Next, we present some of these negative results grouped by the problem used to prove hardness. For an introduction to NP-hardness, see, e.g., [44].

The well-known, weakly NP-complete PARTITION problem [44] has been used in several reductions in boundary labeling. Given a multiset S of positive integers the question is whether we can partition S into two subsets S_1 and S_2 such that $\sum_{x \in S_1} x = \sum_{x \in S_2} x$. Bekos et al. [16] observed that the task of assigning non-uniform labels to two opposite sides of the image region R, say to two vertical sides, is in one-to-one correspondence with the PARTITION problem, if the label heights sum up to twice the height of R. In the context of panorama labelings, Gemsa et al. [48] employ a reduction from the PARTITION problem (that uses the same principle ideas as an earlier reduction of Garrido et al. [46]) to prove that it is NP-hard to determine whether a panorama labeling with vertical leaders exists. A conceptually similar reduction is given by Bekos et al. [14] in the context of labeling collinear sites with non-uniform labels, but directly from the labeling problem by Garrido et al. [46]. In the context of boundary labeling with obstacles, Fink and Suri [39] suggest another reduction from the PARTITION problem to prove that is NP-hard to determine whether a 1-sided labeling with po-leaders and non-uniform labels exists is NP-hard, even in the case of a single obstacle. Finally, a slightly more involved reduction from a variant of the PARTITION problem has been suggested by Bekos et al. [12] to prove that the multi-stack boundary labeling problem with non-uniform labels is NP-hard.

Bekos et al. [10], Lin et al. [70], and Huang et al. [57] proved that different variants of the boundary labeling problem, in which the labels are non-uniform and the objective is the minimization of the total leader length, are NP-hard by employing conceptually similar reductions from an NP-complete scheduling problem commonly known as TOTAL DISCREPANCY [45]. According to this problem, the jobs have different processing times but they all share a common *midtime*, in which the first half of each job must be completed; the goal is to compute a schedule that minimizes the total earliness-tardiness penalty (see also Section 4.2.3).

In the many-to-one boundary labeling setting, Lin et al. [69] prove that the problem of minimizing the number of crossings between leaders can be expressed as a specific instance of the NP-hard TWO-LAYER CROSSING MINIMIZATION problem [35], according to which the goal is to compute a two-layer drawing of a given bipartite graph with minimum number of crossings. Bekos et al. [9] proposed a reduction from the FIXED-LINEAR CROSSING NUMBER problem [76] to prove that the problem of minimizing the number of crossings between leaders is NP-hard even in the case where several leaders to the same label can share a common (horizontal) segment. Recall that the Fixed-Linear Crossing Number problem asks for a linear embedding of a given graph, in which the order of the vertices is fixed along a line L and the edges must be drawn either completely above or completely below L such that the number of edge-crossings is minimum.

4.4   GUIDELINES

Up to now there is no algorithmic technique that prevails over all other techniques. Instead, depending on the concrete problem setting different techniques are suited differently well. Generic techniques such as mathematical programming, dynamic programming, and force-based methods are either slow or they do not give quality guarantees. On the other hand, fast algorithmic techniques with provable guarantees make strong geometric assumptions so that they only can be applied in restricted problem settings. Further, the choice of the best suited labeling technique strongly depends on the quality criteria that should be taken into account. In the following, we summarize for every labeling technique some guidelines when to use it.

1. *Greedy Algorithm.* Greedy strategies are fast, generic and easy to implement. They select for every feature a callout out of a set of candidates. The placement criteria can be expressed both by the construction of the candidates as well as the selection procedure. Typical criteria are the length and slope of the leaders, as well as the importance of the labels. The spatial relation to other callouts (e.g., the distance to the next callout) can only expressed with respect to already selected callouts. In general, as the selection is locally but not globally optimized, greedy strategies give no provable quality guarantees.

2. *Force-Based Algorithm.* Force-based algorithms are fast, generic, and easy to implement. Starting from an initial placement such techniques simulate physical forces to implement placement criteria. The most important criteria such as leader length or distance between callouts can be easily expressed and balanced by the forces. As such approaches typically apply the forces iteratively, they are particularly suitable for interactive systems in which the labeling needs to be adapted based on the changes made by the user. In general, they yield locally optimized solutions but give no guarantees.

3. *Dynamic Programming.* Dynamic programming-based algorithms are generic and exact in the sense that they provide the possibility of optimizing soft constraints while enforcing hard constraints. This comes with the price of high running times and geometric requirements. Further, advanced programming skills are necessary for their implementation. We see their application mainly in static applications in which the quality of the labelings is decisive while the running time of the approaches is of secondary concern.

4. *Weighted Matching.* Reducing a labeling problem to a weighted matching problem can be used for solving instances of boundary labeling where the objective is to minimize the total leader length. As they require fixed positions and unit heights of the labels they are strongly restricted in their applications. As they require a post-processing step to make the computed labelings crossing-free, they are more complicated to implement than, e.g., greedy approaches.

5. *Scheduling.* Labeling techniques based on scheduling are suitable for 1-sided boundary labelings, when the objective is to minimize the total leader length. In contrast to approaches based on weighted matchings they require opo-leaders but also support labels of non-uniform height.

6. *Plane Sweep.* Plane sweep algorithms for external labeling are restricted to boundary labeling with geometric restrictions such as uniform labels or fixed ports. These approaches have asymptotically low running times and they are mostly suitable for the minimization of the total leader length.

7. *Mathematical Programming.* Mathematical programming approaches are based on the selection of candidate callouts. Hence, similar to greedy strategies, quality guarantees can be

expressed by the definition of the candidates as well as the objective function. In contrast to greedy strategies, mathematical programming yields globally optimal solutions. This, however, requires the use of specialized solvers and often leads to high running times. Therefore, they are less applicable in practice, but as they are generic and easy to implement, they can be utilized for small-scale instances and as a tool in time-intensive evaluations of labeling models.

CHAPTER 5

External Labelings with Straight-Line Leaders

Straight-line leaders form the simplest way to establish the visual association between features and their external labels. Further, they can be easily traced by the reader [6]. In this chapter, we first give an overview of different aspects of labelings with straight-line leaders (Section 5.1) and then discuss important contributions (Section 5.2). We note that a survey focusing on external labeling in *medical visualizations* was published by Preim and Oeltze-Jafra [82] in 2014.

5.1 OVERVIEW

Using straight-line leaders for external labeling requires a broad variety of design decisions. As a result, plenty of different approaches have been developed. We summarize the most important properties of these approaches in Table 5.1, which also provides a grouping of several existing algorithms to compute labelings with straight-line leaders. The high-level grouping in the top row of Table 5.1 distinguishes between contour labeling, free labels and non-strict external labelings as introduced in Chapter 2.

- The first group of approaches considers contour labeling assuming that the labels are placed along a pre-defined contour, which means that labels mimic the silhouette of the illustration; see Figure 5.1(a). Especially in professional applications such as atlases of human anatomy domain experts require this property [80]. From an algorithmic point of view, pre-defined contours have the advantage that they reduce the search space by one dimension. On the other hand, this also implies that parts of the labeling region are entirely omitted for label placement, while the chosen contour may not host all labels.

- The second group of approaches relaxes the requirement of pre-defining a contour by freely placing the labels using the entire labeling area; see Figure 5.1(b). Here, one may further distinguish between free label placement within a continuous labeling region, and free label placement within a discretized labeling region.

- The last group of approaches relaxes the requirement that labels may not be placed inside the image region (see Figure 5.1(c)). Note that these approaches still place the labels in the outer regions of the image as best as possible, in contrast to internal la-

Table 5.1: Properties of the approaches for straight-line leaders. References are partitioned into three top-level groups by contour labeling, free label placement, and non-strict external labelings. References considering excentric labeling are marked with "○". (*Continues.*)

	Contour Labeling														Free Labels								Non-Strict										Σ
	[15]	[3]	[25]	[54]	[50]	[27]	[7]	[38]○	[55]○	[62]	[39]	[80]	[6]	[78]	[37]○	[51]	[42]	[94]	[102]	[56]○	[48]	[21]	[53]	[88]	[78]	[84]	[77]	[90]	[5]○	[89]	[79]○	[29]	
S1 labeling contour																																	32
S1.1 rectangle	×	×																					×				× ×			×		×	8
S1.2 circle		×		×		×	×	×	×	×																							6
S1.3 convex hull		×	×	×		×																											6
S1.4 other hull				×	×							×		×											×								3
S2 site placement		×		×	×	×	×					×		×		×	×	×					×		×	×	×	×	×	×	×		12
S3 drawing criteria																																	
S3.1 crossing–free callouts	×	×	×	×	×	×	×	×	×		×	×	×	×		×		×		×	×	×	×	×	×	×	×	×			×		20
S3.2 leader length	×	×	×	×	×	×	×	×	×	×	×	×	×	×	×	×	×	×	×	×	×	×	×	×	×	×	×	×	×	×	×		29
S3.3 label spacing	×	×		×	×	×	×	×	×	×		×	×	×				×		×		×	×	×	×	×	×	×	×	×	×		18
S3.4 leader direction				×	×	×	×	×			×	×	×	×	×			×		×		×	×	×	×	×		×	×	×	×		18
S3.5 vert./horz. aligned labels	×							×		×		×			×	×	×	×							×								12
S3.6 groups/clusters																×	×			×			× ×				×	×					8
S3.7 leader-image occlusion				×	×						×								×				×		×			×					3
S3.8 mixed labeling																																	4
S4 temporal coherence		×		×	×	×								×	×		×			×		×	×	×	×	×	×	×	×	×	×		14
S5 labeling techniques																																	
S5.1 init and improve	×	×	×	×	×	×								×		×				×	×	×	×	×	×	×	×	×	×	×	×		16
S5.2 greedy		×	×	×	×	×								×						×			×		×			× ×		×		13
S5.3 force-based		×		×																×	×		×			×			×	×	×		9
S5.4 dynamic programming								×	×		×	×						×	×														5
S5.5 meta-heuristic																		×	×				×	×									3
S5.6 matching								×																									2
S5.7 plane sweep								×																									1
S5.8 mathematical prog.													×								×	×			×								3

Table 5.1: *(Continued.)* Properties of the approaches for straight-line leaders. References are partitioned into three top-level groups by contour labeling, free label placement, and non-strict external labelings. References considering excentric labeling are marked with "○".

	Contour Labeling														Free Labels								Non-Strict										Σ
	[15]	[3]	[25]	[54]	[50]	[27]	[7]	[38]○	[55]○	[62]	[39]	[80]	[9]	[28]	[37]○	[51]	[42]	[94]	[102]	[56]○	[48]	[21]	[53]	[88]	[78]	[84]	[77]	[90]	[5]○	[89]	[79]○	[29]	32
S6 contributions																																	
S6.1 implementation	×	×	×	×	×	×	×	×	×	×		×	×	×	×	×	×	×	×	×	×	×	×	×	×	×	×	×	×	×	×	×	30
S6.2 analysis: exist. drawings		×		×								×						×			×		×				×						6
S6.3 formal proofs								×	×			×								×	×												6
S6.4 expert interviews												×		×											×							×	5
S6.5 user study													×	×	×							×			×	×			×				7
S7 algorithm type																																	
S7.1 heuristic		×	×	×	×	×	×	×	×			×	×	×	×	×	×	×	×	×		×	×	×	×	×	×	×	×		×	×	27
S7.2 exact	×					×	×		×	×	×										×									×	×		8
S8 community																																	
S8.1 algorithms	×									×	×																						3
S8.2 visual computing		×	×	×	×	×	×	×	×			×	×	×	×	×	×	×	×	×	×	×	×	×	×	×	×	×	×	×	×	×	29

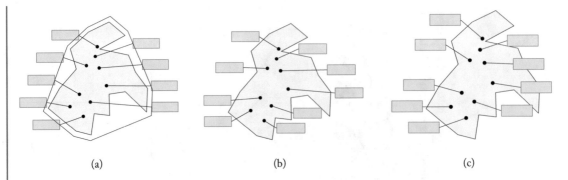

(a)   (b)   (c)

Figure 5.1: Illustration of different types of labeling: (a) contour labeling with a buffered convex hull as contour, (b) free labeling, and (c) non-strict labeling with labels overlapping the image region on the right-hand side.

belings with leaders. So, in a sense they do not fully satisfy the definition of external labeling as given in Chapter 2; we call this placement technique *non-strict* external labeling accordingly.

S1: LABELING CONTOUR

In several applications such as map labeling the labeling contour can be assumed to be given in advance or can be derived from the object to be labeled. In labelings with a pre-defined labeling contour, the shape of the contour has an essential impact both on the visual appearance of the labeling as well as on several algorithmic properties. For boundary labeling, the contour is a rectangle (S1.1), whereas contour labeling uses different possible shapes such as circles (S1.2) and (buffered) convex hulls (S1.3). In rare cases, more complex hulls (S1.4) are also taken into account in order to mimic the shape of the object to be labeled.

S2: SITE PLACEMENT

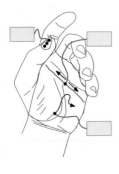

In illustrations of human anatomy the locations of the sites are often pre-defined by a domain expert before the labeling is done [80]. However, for example, in interactive view systems, this is often not adequate, and as a result the positions of the sites need to be calculated. The sites should be placed on salient points of the features such that they do not form dense clusters. Further, considering a 3D object, the projection of the object into the image space is not necessarily connected, but may consist of multiple parts. Hartmann et al. [53] suggested to label the biggest part or the part that promises the shortest leader. Usually, the placement of sites is done in a pre-processing step before the actual labels are placed. Placing sites and labels simultaneously is often more challenging and complicated; see, e.g., [27, 28]. Hence, most of the known techniques can also be installed upstream for the other approaches that require pre-defined sites. Finally, in dynamic scenarios one may also adapt the positions of the sites to improve temporal coherence, e.g., [3, 90].

S3: DRAWING CRITERIA

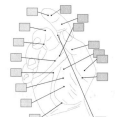

Intersections between callouts can easily distract the reader and let the layout appear cluttered. Hence, special attention is usually paid to guarantee crossing-free callouts (S3.1). Note, however, that intersections may be unavoidable, e.g., if too many labels need to be placed. Similarly, the length of the leaders (S3.2) is considered to be an important drawing criterion to enforce unambiguity; as a result it is quite rare not to explicitly take leader length into account. Another drawing criterion considered in the literature is the spacing between overlapping-free labels (S3.3). This may be achieved, e.g., by penalizing distances between labels [80], by stacking them [3], or by distributing them evenly around the image region [7]. The direction of the leaders (S3.4) is also considered as a drawing criterion, since "uniformly directed leaders" tend to reduce visual clutter. In this regard, several approaches have been introduced, e.g., the angles of the leaders are optimized with respect to some pre-defined directions [27, 94], the leaders radially emanate from a common center [38, 56], or monotonically increasing angles are required when radially ordering the leaders [77, 80]. A vertical and horizontal alignment of the labels or of groups of labels (S3.5) is achieved by stacking the labels horizontally or vertically aligned [48, 51], or by prescribing a rectangular labeling contour [16, 62]. Further, labels can be grouped or clustered [51] (S3.6), and occlusion of the image background by leaders (S3.7) is taken into account [39]. As a special drawing criterion, one may further consider the presence of internal labels. In the positive case, the resulting mixed labelings (S3.8) support both internal and external labels and special care must be paid so to avoid crossings between leaders and internal labels. Note that in the literature one can find many other drawing criteria, depending on the application domain, such as the distance between leaders and sites; for detailed lists of possible criteria, see, for example, [3, 27, 77, 80].

S4: TEMPORAL COHERENCE

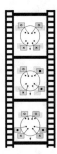

In general, all of the approaches can be used in dynamic scenarios, as each displayed frame of an animation can be labeled independently. However, this easily creates flickering and jumping sites and labels. Therefore techniques that minimize the movement of the sites and labels, e.g., [27, 88, 90] as well as hysteresis techniques that temporally freeze the current labeling [5, 90] have been developed to soften this effect.

S5: LABELING TECHNIQUES

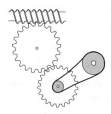

The approaches for obtaining labelings with straight-line leaders are rarely limited to one specific labeling technique, but they use combinations of techniques to obtain an overall strategy. A quite popular scheme first computes an initial labeling and then improves that labeling successively (S5.1). Greedy algorithms (S5.2) and force-based (S5.3) algorithms are easy to implement, run fast and provide the possibility of integrating multiple drawing criteria. Other labeling techniques include dynamic programming (S5.4), meta-heuristics (S5.5), matchings (S5.6) and sweep-line algorithms (S5.7); however, such techniques are less popular. Some approaches use highly specialized algorithms that do not match any of the listed techniques [7, 37, 51, 62]; details are given in Section 5.2.

S6: CONTRIBUTIONS

The contribution of research focusing on straight-line leaders is manifold. Several works describe implementations (S6.1), which are also used for evaluation purposes, e.g., by user studies (S6.5). Such works are usually accompanied with examples generated with the presented approaches. It is also quite common to justify design decisions based on the analysis of existing drawings (S6.2), e.g., atlases of human anatomy and visual dictionaries, and on interviews with domain experts (S6.4) such as surgeons [78], radiologists [77] in the context of medical imaging, designers, and editors working on atlases of human anatomy [80] as well as on interviews with experts of related fields (user interface and graphics design) [28, 29]. On the other hand, fewer works give formal guarantees on certain optimization criteria (S6.3), e.g., the total leader length. Note that apart from analyzing the quality of the labelings, the runtime performance of the presented approaches is also discussed in several works, e.g., [28, 38, 48, 80].

exact

vs.

heuristic

S7: ALGORITHM TYPE

Here, we distinguish between two main types of algorithms that are found in the literature. The first consists of heuristics (S7.1), i.e., approaches that find good but not necessarily optimal solutions. The second contains exact algorithms (S7.2) in the sense that they yield optimal solutions for pre-defined cost-functions.

S8: COMMUNITY

The research on labelings with straight-line leaders stems from the algorithms community (S8.1) and the visual computing community (S8.2). For straight-line leaders, the majority of works stems from the visual computing community, with only few references coming from the algorithms community.

5.2   DETAILED DISCUSSION

In this section we discuss the research on straight-line leaders in detail. To that end, we group the results with respect to the allowed label positions, as, together with the chosen leader type, the position of the labels strongly determines the visual appearance of the labeling.

5.2.1   CONTOUR LABELING

We distinguish three types of contours, which we discuss in the following.

Rectangular contours   Bekos et al. [15, 16] considered rectangles as labeling contours and coined the term *boundary labeling*. For straight-line leaders they considered uniform labels on all four sides of the rectangle and presented algorithms that find for n sites a plane labeling in $\mathcal{O}(n \log n)$ time and a plane labeling minimizing the total leader length in $\mathcal{O}(n^{2+\delta})$ time for any $\delta > 0$; for details refer to Sections 4.2.2 and 4.2.4, respectively. Kindermann et al. [62] considered boundary labeling for straight-line leaders in the context of annotating texts; see Figure 5.2 for an example. They placed the labels either in the left or right margin of the text. For uniform labels they utilized the approach by Bekos et al. [16]. In contrast to opo-leaders running in between the text lines (see Section 6.2.2), straight-line leaders intersect and easily obfuscate the annotated text. Fink and Suri [39] presented dynamic programming approaches that can deal with obstacles, which may not be crossed by leaders. However, their approaches have asymptotically high running times, e.g., for the 1-sided case they obtained a running time of $\mathcal{O}(n^{11})$ and for the 2-sided case they obtained a running time of $\mathcal{O}(n^{27})$ when minimizing the total leader length. Barth et al. [6] used an integer linear programming formulation to create

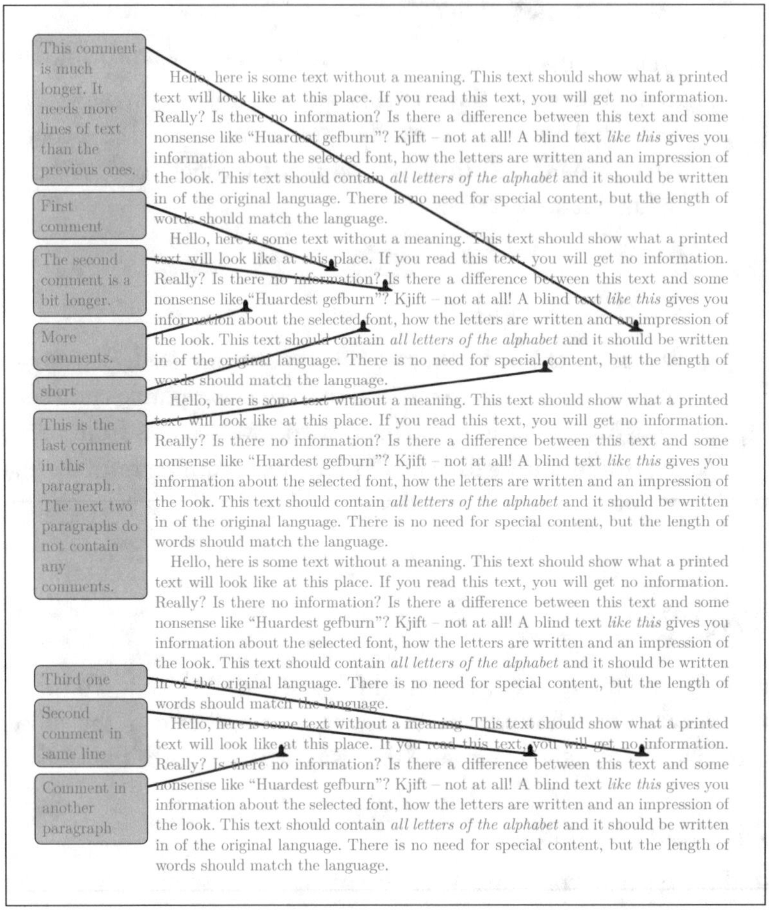

Figure 5.2: Annotating text using straight-line leaders, e.g., in a collaborative text editing process, implemented in the LaTeX luatodonotes package [62].

1-sided boundary labelings with straight-line leaders (Section 4.2.5). They used these labelings in a user study on the readability of different leader types. They particularly showed that straight-line leaders perform well when the user is asked to associate labels with their sites and vise versa. However, concerning the aesthetic preferences the users favored non straight-line leaders, in particular, po-leaders and do-leaders.

Circular contours   Ali et al. [3] suggested circular contours and contours that mimic the silhouette of the illustration (e.g., by a buffered convex hull; see Figure 5.3 for an illustration). In an initial procedure the labels are placed along these contours such that they are stacked and no label-label overlaps occur. Afterward, the positions of the labels are switched until all leader-

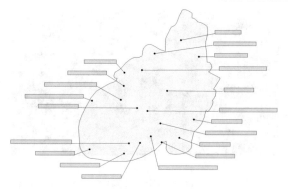

Figure 5.3: A labeling with straight-line leaders and labels mimicking the silhouette of the illustration produced adopting the approach by Ali et al. [3].

leader intersections are resolved. In a final compaction step, the requirement of being placed on a contour is relaxed and the labels are pushed toward the illustration to reduce the lengths of the leaders. In a companion paper, Götzelmann et al. [49] described a system for labeling 3D models with internal and external labels. In this system the external label placement is done by the approach by Ali et al. [3]. Hartmann et al. [54] presented a high-level description of that system.

Battersby et al. [7] proposed *ring maps* considering circles as contours. Uniform labels are evenly distributed around the circle such that each label is oriented with the ray that emanates from the center of the circle and goes through the center of the label. Hence, the labels are not axis-aligned but radially aligned. The order of the labels around the circle is chosen such that the resulting labeling is plane. Fink et al. [38] also considered circles as contours in the context of excentric labeling. In contrast to related work [37, 56], the labels are required to touch the circle. For all presented algorithms they gave formal guarantees on the drawing criteria. In particular they presented dynamic programming approaches that support general weighting functions. Haunert and Hermes [55] considered a closely related setting using radial leaders but horizontally aligned labels (see, e.g., Figure 5.4 for an illustration). They reduced the problem to finding maximum weight independent sets in conflict graphs of the labels, which they solved by means of dynamic programming.

Convex hulls and silhouette contours   Similar to Ali et al. [3], Čmolík and Bittner [27] considered rectangles, circles, and convex hulls as pre-defined labeling contours (see, e.g., Figure 5.5 for an illustration). For each area to be labeled they generated a set of candidate callouts that differ in the chosen sites and the chosen position of the labels on the contour. Using a rating of the candidates based on fuzzy logic, they utilized a greedy algorithm that selects a candidate for each feature. In a successive work, Čmolík and Bittner [28] adapted these techniques to label *ghosted views* of 3D models, i.e., views that use transparency to show occluded parts of the

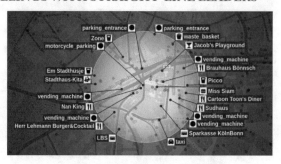

Figure 5.4: A labeling created with the approach presented by Haunert and Hermes [55].

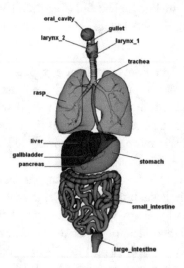

Figure 5.5: A labeling of a 3D model in which the labels are arranged on the boundary of a predefined labeling contour [28].

models. They particularly focused on the placement of the sites on salient parts of the model. They evaluated the approach in an extensive user study. Similar to the study by Barth et al. [6], the users were asked to associate labels with their highlighted features. The stimuli were created by their approach as well as by experts. Čmolík et al. [29] integrated a modified version of the algorithm by Čmolík and Bittner [28] into an algorithm that supports mixed labelings. In particular, the modification supports non-convex shapes as labeling contours.

Bruckner and Gröller [25] presented an approach for labeling 3D models using a simple iterative algorithm that places the labels along the convex hull of the projected model. Overlaps are resolved by moving the labels along the contour and by exchanging the positions of overlapping labels. If not all overlaps can be resolved, a greedy algorithm selects the labels by priority

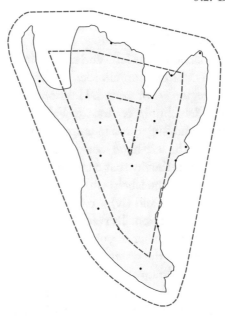

Figure 5.6: Hulls generated by the approach by Niedermann et al. [80]. The outer hull is used for the placement of the labels. The two inner hulls are used to restrict the leader candidates: a leader may only intersect a hull if its site is contained in that hull.

excluding overlapping labels. Intersections between callouts are resolved in a post-processing step based on the approach presented by Bekos et al. [16]. Niedermann et al. [80] used convex hulls to prescribe the labeling contours of illustrations in atlases of human anatomy; see Figure 5.6 for an example. They presented a general dynamic programming approach that labels a given set of sites with respect to a set of hard and soft constraints. They proved that the calculated solution respects all hard constraints and is optimal with respect to all soft constraints. In contrast to other dynamic programming approaches, the cost-function not only supports the rating of single leaders, but also of consecutive leaders. Thus, they could incorporate drawing criteria such as distances between labels and monotonically increasing angles between leaders.

Götzelmann et al. [50] considered external labeling in interactive 3D visualizations. They used the *orbit* of a figure as contour, i.e., a hull around the figure such that each point on the hull has the same distance to the figure. They computed the initial positions of the labels using a simple greedy algorithm. The movement of the labels and the sites is controlled by a multi-agent system such that for each label there are agents, i.e., local strategies, that control the behavior of the labels. They paid special attention to temporal coherence and combined their approach with internal label placement.

5.2.2   FREE LABEL PLACEMENT

Fekete and Plaisant [37] introduced *excentric labeling*. For straight-line leaders they presented a simple approach that stacks the labels on the left- and right-hand side of the focus region either according to their vertical or to their horizontal order. To improve temporal coherence, labels are only placed when the focus region is not moved. Heinsohn et al. [56] also considered excentric labeling for straight line leaders, but in contrast to Fekete and Plaisant [37] the labels are displayed all the time. Heinsohn et al. suggested four approaches to place labels: (i) an approach that places all labels on a stack on the left-hand side of the focus region; (ii) a radial approach, i.e., the leader of each label is part of the ray that emanates from the center of the focus region and goes through the point feature of the labels; (iii) a force-based approach that prefers labels with radial leaders (see Section 4.1.2); and (iv) a *cake-cutting* approach that places the labels equally distributed around the focus region. In case that not all overlaps can be resolved, the focus region is shrunk to reduce the number of point features within the focus region.

Götzelmann et al. [51] extended the system by Hartmann et al. [54] by supporting grouping of labels. More precisely, for each group of features they first compute the sites of the features and determine the centroid of those sites. For each centroid they create rectangle candidates in the labeling region that can host the grouped labels. Within a group the labels are vertically aligned. Vollick et al. [94] formalized the external labeling problem as an energy function that is composed by nonlinear terms each representing one drawing criterion. The exact weighting of the terms was learned from existing visualizations (e.g., from hand-made drawings) by applying *nonlinear inverse optimization*. The authors used simulated annealing to find a locally optimal labeling from the energy function.

While those approaches make use of a continuous labeling region, the following discretize the labeling region to find appropriate label positions. Fuchs et al. [42] partitioned the labeling region into small rectangular regions. Using a simple greedy algorithm the labels were assigned to these regions such that the length of the leaders is minimized. Apart from labeling all features of the illustration, they also considered a *labeling lens*, which is conceptually identical to the focus region considered in excentric labeling. In this case the same approach was applied but restricted to that region. Further, the approach was also used for more general settings, where the image region consisted of multiple parts. Wu et al. [102] used external labeling to annotate a metro map with station names and photos; see Figure 5.7. To that end, they underlied the metro map with a fine raster describing the free and occupied regions of the map. Only cells that are not occupied can be used for the label placement. They applied a genetic algorithm combined with a greedy algorithm to determine the position of the labels. In particular, they placed textual labels and image labels independently allowing label-leader intersections between the two groups. Gemsa et al. [48] considered the special case of panorama labeling, e.g., for labeling the skyline of a city or the peaks of mountains; see Figure 5.8(a). They showed that in the proposed model it suffices to consider a finite set of candidates per feature to cover the optimal intersection-free labeling. Further, they presented a dynamic programming approach and

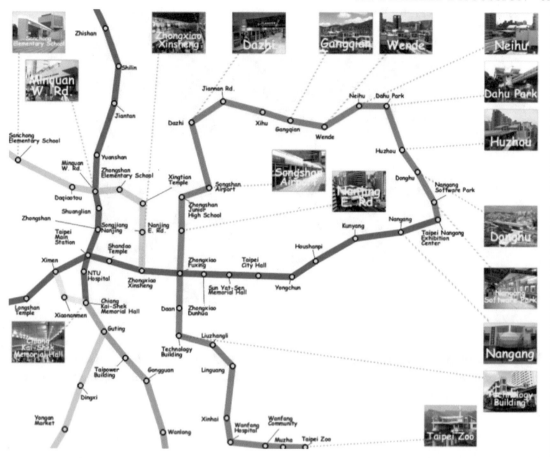

Figure 5.7: A sample external labeling used to annotate a metro map with station names and photos [102].

mixed integer linear programming formulations for computing optimal labelings with respect to different cost functions and settings. Bobak et al. [21] also considered panorama labeling, but for a 3D model whose perspective is changed over time; see Figure 5.8(b). They present a mathematical programming formulation for both the offline and online setting. In the online setting, in which the sequence of frames is not known in advance, they compute the labeling of a frame based on the preceeding frame.

5.2.3   NON-STRICT EXTERNAL LABELING

Hartmann et al. [53] presented an approach for labeling 3D objects. As they considered complex geometric features, their approach first determines the location of the sites by shrinking

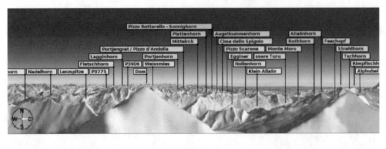

(a)

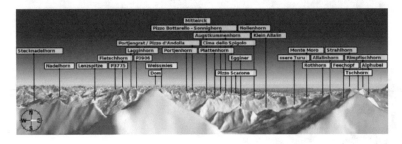

(b)

Figure 5.8: Panorama labelings (a) created with the approach by Gemsa et al. [48] minimizing the number of rows and (b) created with the approach by Bobak et al. [21] optimizing the labeling over a given sequence of frames.

the feature until a single point remains. They argued that this point is an appropriate choice of a non-convex object, because it is "placed at a visual dominant region." For the actual labeling they presented an approach that consists of two phases. In the first phase, an initial layout is computed using a force-based approach (see Section 4.1.2). In the second phase, possible overlaps between labels are minimized using a combination of the same force-based approach and a greedy algorithm. While an external labeling is preferred, labels may overlap the image region. Tatzgern et al. [90] adapted the approach by Hartmann et al. [53] to label 3D explosion diagrams. They first computed an initial layout by forming clusters of labels with similar texts, and placing each label using the first phase of the approach presented by Hartmann et al. [53]. Then, in a second step they selected for each cluster its best label with respect to a set of optimization criteria. Afterward, they applied the second phase of Hartmann et al. [53] to minimize overlaps. They applied their algorithm in a dynamic 3D scene. However, they did not enforce a strict separation between labeling and image region to the effect that the labels could overlap the labeled 3D object. To ensure temporal coherence, the order of the labels and sites is fixed during the movement of the camera. When the camera stops, possible intersections are resolved.

Stein and Décoret [88] formalized the external labeling problem of 3D models in the 2D projection space as an energy function whose minimum yields a labeling with short leaders, crossing-free callouts, and labels that avoid the image region. They used a greedy strategy to find a locally optimal labeling. They improved temporal coherence by penalizing positions of labels that lie too far away from the previous positions. They also avoided flickering effects that occur when sites become invisible during interaction.

In the context of excentric labeling, Balata et al. [5] presented an approach for labeling moving unmanned aerial vehicles. They did not explicitly distinguish between image and labeling region, but they used a force-based approach that separates the labels from their sites. This approach ensures that leaders do not intersect, but labels may overlap each other. To improve temporal coherence, they considered (a) forces that soften distracting flickering effects, and (b) the possibility of temporally freezing labels. In a user study they showed that freezing labels led to lower error rates of the users than the other approach.

Mühler and Preim [78] and Mogalle et al. [77] presented two approaches of non-strict external labeling with labeling contours. If necessary (e.g., due to missing free space), Mühler and Preim [78] relaxed the requirement on placing labels along a contour, but allowed the labels to overlap unimportant structures of the image. To that end, they extended the approach of Ali et al. [3] such that it also supports the annotation of ghost views used in surgical planning. Further, they conducted a user study in which the participants were asked to judge 24 images using different drawing styles. Among others, the user study showed that the participants preferred grouped labels. Mogalle et al. [77] placed the labels inside of the image boundary such that they touch the boundary. They presented both a greedy algorithm that selects labels candidates and an approach that shifts the labels along the contours. In order to avoid overlaps with other pictorial elements (e.g., legends) they relaxed the requirement that the labels are placed along a rectangle. Instead, they integrated pictorial elements into the shape of the contour.

Finally, Niedermann and Haunert [79] presented an excentric labeling approach that allows the user to interactively move a lens on top of a map; see Figure 5.9. The point features contained in the lens are labeled using radial leaders. The labels are positioned depending on the distance of the point features to the center of the lens. To that end, they utilize a fish-eye projection, which also sustains temporal coherence. In order to reduce clutter produced by leaders running closely in parallel, they bundle such leaders. In an additional optimization step based on a force-based method, they reduce label overlaps.

5.3   GUIDELINES

Straight-line leaders are the shortest and most natural connection between a site and its label. They can be followed by the eye easily, without any bends or detours. Hence, they are widely used in information graphics and visualization applications. However, if done carelessly, they bear the risk of creating irregular and cluttered slope patterns that might spoil an illustration. We

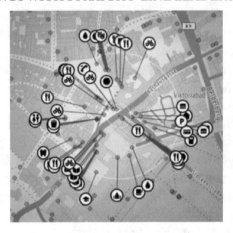

Figure 5.9: A fish-eye labeling produced by the approach of Niedermann and Haunert [79].

recommend to consider the following aspects when choosing to implement or apply a straight-line labeling method and refer to Table 5.1 as a selection aid.

1. Determine the labeling contour: is a rectangular bounding box or a bounding circle appropriate? Such contours form a well-structured and simple separation between image region and labeling region and one can choose from a large variety of labeling algorithms. Otherwise, use a suitable fitted convex hull or a more fine-grained silhouette as the contour, but avoid overly complex shapes.

2. If the number of labels is a good match to the length of the contour, then using a contour-based labeling model will usually result in a clean labeling; however, if labels need to be packed tightly or the contour is too short to host all labels, then choose a more flexible free or non-strict labeling model. Also, if the setting is dynamic with interaction and immersion, the larger freedom in placement options of non-strict models might be an advantage.

3. Collect the visual requirements and drawing criteria needed. In almost all cases, we recommend to choose crossing-free callouts and to include leader length minimization so that the site-label distance is small. If the application permits it, restricting the leader direction (e.g., radial leaders, vertical/horizontal leaders, or monotonically increasing leader slopes) results in cleaner and less intrusive labelings. For more advanced requirements, use Table 5.1 for your orientation. If none of the existing methods meet your needs, then we recommend easily adaptable and configurable approaches that use generic weight or cost functions for prioritizing callout candidates or unconstrained quality functions for the whole labeling.

4. Collect the computational performance requirements. Are low running times a high priority or are even real-time algorithms needed? Then fast heuristics, especially those doing

iterative improvements (init and improve), using force-based models or applying greedy optimization are usually the method of choice. If, however, instances are small or layout optimality is most critical, then exact methods are recommended.

CHAPTER 6

External Labelings with Polyline Leaders

Polyline leaders form a natural generalization of straight-line leaders by supporting bends. To facilitate readability, polyline leaders usually have a schematic and tidy appearance caused by the restriction to segment orientations that are aligned with the Cartesian coordinate axes or possibly with their two bisecting diagonals. As one would expect, the most natural requirement in labelings with polyline leaders is the avoidance of crossings between leaders. However, bend minimization is another criterion which becomes important, as has also been observed in the early survey on boundary labeling by Kaufmann in 2009 [60].

In Section 6.1, we give an overview of the existing literature on labeling with polyline leaders in terms of several characteristic properties and grouped by different types of input specifications. It is worthwhile to mention that the majority of the existing techniques considers the boundary labeling setting, where the contour C_A is a rectangle. Section 6.2 provides a more detailed discussion of important contributions.

6.1 OVERVIEW

Table 6.1 provides a grouping of several existing algorithms to compute labelings with polyline-leader. The grouping is based on input specifications. In the table, we have also identified 11 important properties P.1–P.11 (briefly introduced and discussed in the following) that provide a further classification each of the existing algorithms. The high-level grouping as seen in the top row of Table 6.1 first considers four types of static external, mostly boundary labeling, problems, while the latter two extend to dynamic and many-to-one problems, respectively. More precisely:

- The first group focuses on *point feature labeling*, some of which assume the points to be word positions in line-based text documents.

- While in the former case, labelings are usually unconstrained in the background and feature layers, *text labeling* often requires leaders that run horizontally between two text lines and may use a track-routing area in a narrow strip between text and labels to move vertically.

- A natural generalization of point feature labeling is *labeling of area features*, in which each leader's site within the feature can still be chosen freely by the labeling algorithm.

Table 6.1: Properties of the approaches for polyline-line leaders. References are partitioned into six top-level groups by feature type (*points*, *text*, *areas*), use of internal and external labels (*mixed*), features changing over time (*dynamic*), and many-to-one labeling (*X-to-1*). References considering excentric labeling are marked with "○". (Continues.)

	Points											Text			Areas			Mixed			Dynamic				X-to-1			Σ
	[16]	[12]	[18]	[14]	[10]	[57]	[63]	[22]	[6]	[23]	[47]	[71]	[62]	[70]	[13]	[39]	[105]	[72]	[11]	[73]	[37]○	[3]	[19]○	[8]	[69]	[89]	[9]	27
P1 leader type																												
P1.1 po-leaders	×	×	×	×		×	×	×	×	×	×	×	×	×	×	×		×				×		×	×	×	×	17
P1.2 opo-leaders	×	×	×			×	×	×	×			×	×	×	×	×			×	×					×	×		14
P1.3 do- and pd-leaders			×		×				×	×						×	×						×					6
P1.4 other		×	×				×		×		×	×	×			×				×								3
P2 objective																												
P2.1 length minimization	×	×	×	×	×	×	×	×	×		×	×	×	×	×	×		×	×	×		×		×	×	×	×	20
P2.2 bend minimization	×	×	×		×	×	×	×	×						×	×							×					6
P2.3 multicriteria																					×				×			5
P2.4 other		×		×			×				×	×	×				×								×	×	×	11
P3 contour																												
P3.1 boundary 1-sided	×	×	×		×	×	×	×	×		×	×	×	×	×	×	×	×	×	×		×		×	×	×	×	20
P3.2 boundary 2-sided (opposite)	×	×	×			×	×	×				×	×		×	×		×	×			×		×	×	×	×	14
P3.3 boundary 2-sided (adjacent)							×	×		×																		3
P3.4 boundary 3-sided							×	×		×																		3
P3.5 boundary 4-sided							×	×		×					×					×	×		×		×			7
P3.6 non-rectangular	×				×															×	×							4
P4 label type																												
P4.1 uniform labels	×	×	×	×	×	×	×	×	×	×	×	×	×	×	×	×	×	×	×	×	×	×	×	×	×	×	×	23
P4.2 non-uniform labels	×	×	×			×	×	×	×	×			×		×	×		×	×		×	×		×	×	×	×	12
P5 label positions																												
P5.1 fixed positions	×		×	×		×	×	×	×	×	×	×	×	×	×	×	×	×	×		×	×	×	×	×	×		14
P5.2 sliding positions		×	×		×	×		×			×	×			×				×							×	×	14
P6 label ports																												
P6.1 fixed ports	×	×	×	×	×	×	×	×	×	×	×	×	×	×	×	×	×	×	×	×	×	×	×	×	×	×	×	25
P6.2 sliding ports	×	×	×			×		×	×		×				×	×		×								×		13

Table 6.1: *(Continued.)* Properties of the approaches for polyline-line leaders. References are partitioned into six top-level groups by feature type (*points*, *text*, *areas*), use of internal and external labels (*mixed*), features changing over time (*dynamic*), and many-to-one labeling (*X-to-1*). References considering excentric labeling are marked with "○".

	Points											Text			Areas			Mixed			Dynamic				X-to-1			Σ
	[16]	[12]	[18]	[14]	[10]	[57]	[63]	[22]	[6]	[23]	[47]	[71]	[62]	[70]	[13]	[39]	[105]	[72]	[11]	[73]	[37]○	[3]	[19]○	[81]	[69]	[89]	[6]	27
P7 leader obstacles								×				×	×			×		×	×	×								7
P8 contributions																												
P8.1 formal proofs	×	×	×		×	×	×	×		×	×	×	×	×	×	×		×	×	×	×	×	×	×	×	×	×	21
P8.2 implementation	×	×	×	×		×			×		×		×		×	×	×	×	×		×	×	×	×	×	×		15
P8.3 user study									×		×						×											4
P9 algorithm type																												
P9.1 exact	×	×	×	×	×	×	×	×	×	×	×	×	×		×	×		×	×	×				×	×	×	×	23
P9.2 approximation																			×							×		2
P9.3 heuristic	×							×	×		×	×					×			×				×	×			7
P10 techniques																												
P10.1 dynamic programming	×	×	×		×	×	×	×		×	×	×	×	×	×	×		×	×	×				×	×		×	14
P10.2 plane sweep	×	×	×									×	×	×		×		×	×		×			×		×		9
P10.3 weighted matching	×			×	×																				×	×	×	5
P10.4 scheduling						×																						4
P10.5 init and improve									×																			3
P10.6 mathematical programming																	×	×	×		×	×	×					4
P10.7 meta heuristics											×								×						×	×	×	3
P10.8 greedy							×	×											×									1
P10.9 other algorithm	×				×											×			×								×	4
P10.10 NP-hardness	×	×	×	×	×	×	×	×	×	×	×			×			×	×	×	×				×	×	×	×	9
P11 community																												
P11.1 algorithms	×	×	×	×	×	×	×	×	×	×	×	×	×	×	×	×	×	×	×	×	×	×	×	×	×	×	×	18
P11.2 visual computing	×	×							×		×						×			×		×	×	×		×	×	9

- The next group considers *mixed labeling*, which combines the placement of internal labels with external labels. Here internal labels form obstacles that restrict the feasible positions of the leaders.

- The group of dynamic labeling algorithms take into account point features that move over time as the user may shift a focus region or change the zoom level. While a static labeling algorithm usually does not preserve temporal coherence when applied to dynamic inputs, an algorithm for a dynamic labeling problem, conversely, can easily be applied to a static input.

- The last group considers many-to-one (X-to-1) labelings, where a single label refers to multiple features.

In the rest of this section, we introduce and briefly discuss the classification done by the parameters P1–P11.

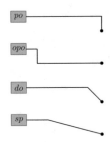

P1: LEADER TYPE

One of the primary properties to distinguish different labeling algorithms is the used leader type. In particular, most of the existing algorithms to compute labelings with polyline leaders focus on leaders with orthogonal segments. Such leaders are usually either of type po (P1.1) with a single bend inside the image region or of type opo (P1.2) with two bends in a narrow *track routing area* outside the image region but inside the contour C_A. A natural generalization is formed by the orthodiagonal leaders, which are usually of type do or pd with a turning angle of $\pm 45°$ (P1.3). These leaders are shorter than the respective po-leaders and have been empirically confirmed to be aesthetically more pleasing, yet slightly less readable than po-leaders [6]. Finally, it is worth mentioning that there exist algorithms in the literature for producing labelings with polyline leaders that do not use fixed slopes, but rather include segments of type s between the site, the bends, and the label (P1.4).

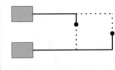

P2: OBJECTIVE

The most common objective functions (see also Section 3.2) studied in the literature for labelings with polyline leaders are the total leader length (P2.1) and the total number of bends (P2.2). A small leader length implies a short distance between site and label and thus optimizes the locality of labels. It also minimizes the ink used to plot the leaders, and thus the amount of overplotting of the image region. Note that a leader without bends is usually also a shortest possible leader, whereas for leaders with bends no such claim about their length can be made. There exist also methods that use a multicriteria objective function (P2.3) taking into account, e.g., leader lengths, but also separation between callouts and displacement of features or labels [3, 19, 37, 105]. Other objective functions (P2.4) are either arbitrary quality measures based on a single leader [18], finding a feasible labeling at all [71], maximizing the number of labeled features [63], maximizing the label size [12], maximizing the number of internal labels in a mixed model [11, 72, 73], or minimizing the number of leaders in many-to-one labeling [9, 68, 69].

P3: CONTOUR

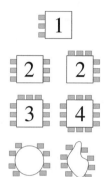

A quite common approach to produce labelings with polyline leaders is by adopting the boundary labeling model, where the contour is the bounding box of the image region R. Therefore, a natural distinction among the different algorithms is based on the number of sides of R that can be used for attaching labels. The simplest and most-studied case is 1-sided boundary labeling (P3.1). From an algorithmic point of view the 1-sided problem reduces to finding the best ordering of the labels on that side. This is a less complex situation than multi-sided labelings, where one needs to determine a side assignment together with an ordering for the labels on each side. Interestingly, 2-sided settings having labels on two opposite sides (P3.2), say left and right of R, are well studied and admit more efficient algorithms than the case of labels on two adjacent sides (P3.3) of R, say top and right, as we will shortly see. The same also applies to 3-sided labelings (P3.4). In applications with labels being names or other short text strings, using the two opposite sides left and right of the image region is more natural than using two adjacent sides. This due to the fact that the height of a horizontal text label is small and thus many such labels can fit. The top and bottom sides, however, offer space for much fewer labels as the widths of the labels quickly add up to the image width, unless the labels are rotated by 90°. Only if the labels have aspect ratio closer to 1, e.g., for icons or acronyms, or are rotated accordingly, then all four sides (P3.5) are equally well suited for placing labels. The study of labelings with polyline leaders is not limited to the case of a rectangular contour. There exist also variants with different contours (P3.6) like circles in excentric labeling [19, 37], convex polygons [3, 73], or arbitrary silhouettes [3].

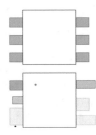

P4: LABEL TYPE

One may distinguish between two different types of labels; Uniform labels (P4.1) and non-uniform labels (P4.2). In the former case, all labels have the same size (or at least the same height as found in single-line text labels) and thus can be represented as a set of empty boxes distributed along the contour. Algorithmically speaking, one has to find a matching that assigns each feature to one of these empty boxes; then, the label text is guaranteed to fit into the box. The more general case of non-uniform labels is usually more difficult to solve as the placement of the labels now depends on the ordering of the leaders around the contour, e.g., imagine a large label consuming more space and pushing neighboring labels aside. Note that many optimization problems become NP-hard in the case of non-uniform labels due to the close relation to job scheduling problems (see Section 4.3). Still, for a number of problem settings, efficient algorithms exist even for non-uniform labels [57, 62, 70, 71, 81].

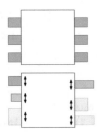

P5: LABEL POSITIONS

The label positions are either fixed on the contour (P5.1) or they can slide along it (P5.2). The former case is more restrictive but also simplifies the algorithmic problem in many cases as one degree of freedom is removed. Fixed positions are mostly used for uniform labels (P4.1), since for non-uniform, individually sized labels fixed positions already dictate the assignment of features to labels. In combination with property P6.1 (ports at fixed positions), fixed-position, labels give rise to leaders whose geometry is fully specified once a feature is assigned to a label. On the other hand, in some situations sliding labels are needed to compute feasible labelings at all, e.g., when labels of non-uniform size can be stacked or when leaders are blocked from reaching certain label positions due to obstacles.

sliding   fixed
port   port

P6: LABEL PORTS

Label ports can either be at fixed positions (P6.1) or they can slide along one side of the label box (P6.2). Again, fixed port positions (in combination with fixed reference points (P5.1)) are algorithmically often easier to handle because the geometry of leaders is fully determined by the position of the site and the position of the fixed port. Moreover, fixed ports, e.g., centered at the label, often produce tidier labelings as in consequence the distance between neighboring leaders is uniform. Sliding ports may produce shorter leaders, e.g., when connecting a site that sits below a label to the bottom corner of its label rather than to its midpoint. Many algorithms even work for both types of ports.

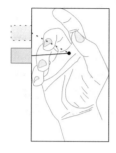

P7: LEADER OBSTACLES

Obstacles are areas in the image region that cannot be crossed by a leader. These can be important parts of the image, individual words in a text document, or internal labels in case of mixed labeling. Obstacles are usually modeled as polygons. They restrict the options of connecting a feature to a label position and thus increase the difficulty of the algorithmic problem, e.g., they can make an instance infeasible. A special situation arises in mixed labeling [11, 72, 73], where the algorithm can decide whether an internal label, i.e., an obstacle, is placed.

P8: CONTRIBUTIONS

Here, we distinguish three (non-exclusive) types of results that are often found in the literature on labelings with polyline leaders. The first contains exact algorithms (see also property P9.1) accompanied with formal proofs of correctness, optimality of the solution, and asymptotic running times (P8.1). Not all of these algorithms have been implemented and some even have running times that are too high to be of practical relevance. On the positive side, they provide certificates that the corresponding problems are polynomial-time tractable. Besides these algorithms, one can find formally analyzed algorithms together with proof-of-concept implementations, as well as works that do not contain formal results and instead report about implemented systems and algorithms for polyline leaders that are justified by case studies or experiments (P8.2). Finally, it is worth noting that there also exist user studies on the computed labelings (P8.3).

exact

approx.

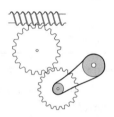

heuristic

P9: ALGORITHM TYPE

Here we consider three different types of algorithms found in the literature. As expected, several algorithms in the literature are exact in the sense that they provide mathematically optimal solutions to their respective labeling problems (P9.1). Approximation algorithms (P9.2) cannot guarantee to find optimal solutions, but they are accompanied by formal proofs that provide an approximation guarantee that quantifies how far a solution can be from the optimum in the worst case. Finally, heuristics (P9.3) may compute good and sometimes optimal solutions, yet no formal proofs and guarantees are provided in the respective papers.

P10: TECHNIQUES

This property refers to the algorithmic techniques and principles used in the different papers. The main features and some generic descriptions of these techniques are sketched in more detail in Chapter 4. Dynamic programming (P10.1) is a quite common technique to compute labelings with polyline leaders. It is based on subdividing a problem instance into two (or more) independent subinstances and then creating a solution from compatible subsolutions. Especially 1-sided boundary labeling problems can also be solved by sweep-line algorithms (P10.2) that visit the sites and labels in a suitable sequence, say top to bottom, and compute the leaders at discrete events, e.g., when new points or labels are reached. The weighted matching technique (P10.3) is well suited for problems optimizing over a finite set of possible site-label assignments modeled as a weighted bipartite graph, e.g., for instances with fixed label positions, fixed ports, and uniform labels. A scheduling algorithm (P10.4) is used for the 1-sided labeling of collinear points [14], while the scheduling technique is also the basis for a number of NP-hardness reductions [10, 57, 70]. There exist also works that use global optimization techniques such as SAT solving [11] and mathematical programming (P10.6), which includes integer linear programming [6, 11] and quadratic programming [105]. Meta heuristics (P10.7) include genetic algorithms [71] and simulated annealing [68] or heuristics that first construct an initial solution and then improve it by local modifications (P10.5) [3, 19, 37]. A greedy algorithm (P10.8) is used for a crossing-minimization heuristic in many-to-one labeling [69]. Among other techniques (P10.9) are algorithms that resolve leader crossings by swapping some site-label assignments that do not affect the sum of leader lengths [16], recursive approaches [11], and a reduction to finding weighted independent sets in outerstring graphs [22]. Of course, not all optimization problems are polynomial-time tractable, and therefore several NP-hardness reductions exist (P10.10, see also Section 4.3), mostly for labeling problems with non-uniform labels. These are used to justify heuristics and approximation algorithms or simplifications of the model, e.g., the use of uniform labels.

P11: COMMUNITY

This property is identical to S8 for straight-line leaders. Interestingly, poly-line leaders are studied primarily in the algorithms community (P11.1), but they are also gound in several papers in the visual computing community (P11.2).

6.2   DETAILED DISCUSSION

We start our discussion with two types of orthogonal leaders, namely po-leaders (Section 6.2.1) and opo-leaders (Section 6.2.2). Then we discuss 1-bend orthodiagonal leaders (Section 6.2.3). Since the vast majority of the literature considers rectangular contours, i.e., the boundary labeling problem, we assume that the contour C_A is an axis-aligned rectangle, unless stated otherwise.

6.2.1   po-LEADERS

Orthogonal po-leaders are the simplest and most commonly used polyline leader type in external labeling. They are frequently found in the algorithmic literature, but also appear in professional information graphics and are implemented in some practical labeling methods.

1-Sided Boundary Labeling
In the 1-sided setting, one of the simplest boundary labeling problems is to assign external labels of uniform size with fixed ports and thus fixed reference points aligned on a single, say vertical, side of the image region R to a set of n point features in R such that the po-leaders are crossing-free. The most commonly used objective function is to minimize the total leader length. For this problem Bekos et al. [16] presented an $\mathcal{O}(n^2)$-time algorithm to find a length-minimal solution. Their algorithm runs in two phases. First, sites and labels are matched so that they have the same vertical order. Second, leader crossings are iteratively resolved without changing the total leader length. Later, Benkert et al. [18] presented an algorithm based on the sweep-line paradigm that visits all sites and labels in their vertical order. Whenever a new label becomes available, a site is selected such that its leader is guaranteed to be crossing-free and the total length remains minimum. Their algorithm improved the running time from $\mathcal{O}(n^2)$ to $\mathcal{O}(n \log n)$.

A very different method to compute length-minimal solutions is a simple integer linear program (ILP) proposed by Barth et al. [6]. This approach generalizes to arbitrary leader shapes

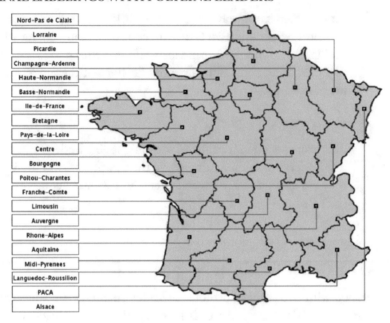

Figure 6.1: An example of a 1-sided boundary labeling with type-po leaders taken from [18].

and label positions, as it basically computes a minimum-length crossing-free perfect matching between the given sites and fixed label positions. While in general ILP solving does not scale well, it is still a practical approach for smaller instances. Nöllenburg et al. [81] relaxed the restriction to fixed reference points with uniform labels and allowed sliding reference points and non-uniform labels. They presented another sweep-line algorithm that computes a length-minimal and crossing-free solution for this problems in $\mathcal{O}(n \log n)$ time. Huang et al. [57] additionally relaxed the constraint to have fixed-port labels and allowed sliding ports. In this model, they proposed an $\mathcal{O}(n^3)$-time algorithm for minimizing the total leader length.

A heuristic framework for external labeling has been proposed and implemented by Ali et al. [3]. They considered rectangular contours as well as other silhouettes and used rectilinear leaders as one option. The algorithm first determines suitable label positions, e.g., by projecting each site to the closest point of the contour or by using force-based methods, and then tries to iteratively resolve leader crossings until a crossing-free solution is found. While this can produce reasonable labelings, no guarantees on optimality or termination are given.

Fink and Suri [39] proposed a collection of dynamic programming algorithms for boundary labeling with *obstacles*. Each obstacle is modeled as a rectilinear polygon and no leader may cross any obstacle. With their algorithm for 1-sided boundary labeling with po-leaders and uniform labels, they could minimize the total leader length in $\mathcal{O}(n^4)$ time if label positions are fixed and in $\mathcal{O}(n^7)$ time otherwise. Their method also extends from point to area features at the cost of another linear factor. If, however, the labels have non-uniform size, they proved NP-hardness for deciding the existence of a crossing-free solution. A more restricted case of boundary labeling with obstacles has been studied by Löffler and Nöllenburg [72]. They made the simplifying assumptions that all obstacles are congruent rectangles (which may model a set of internal labels). They provided dynamic programming algorithms for placing non-crossing po-leaders such that the number of intersected obstacles is minimized. For some cases, e.g., when the obstacles can intersect each other or leaders may cross, they also proved NP-hardness results.

For objective functions other than length minimization, Benkert et al. [18] proposed a general $\mathcal{O}(n^3)$-time algorithm using a dynamic programming approach that optimizes the labeling using uniform, fixed-position labels with sliding ports. Kindermann et al. [62] generalized this algorithm to instances with non-uniform labels using rasterized label positions, where one label may occupy multiple raster slots. This increases the running time to $\mathcal{O}(n^4 m^3)$ and the space consumption to $\mathcal{O}(n^3 m^2)$, with $m > n$ being the number of raster slots. Kindermann et al. used this algorithm for text labeling and made some adaptations to preferably route the p-segments between two lines of text rather than striking them out. Another interesting variation is 1-sided multi-stack labeling [12], where labels take positions in two or more label stacks on the same side of C_A. However, using po-leaders the authors only showed that for non-uniform labels it is NP-hard to decide the existence of a crossing-free two-stack labeling. Feasible algorithms in this model, e.g., for uniform labels, are missing.

Finally, for the many-to-one boundary labeling, where sets of point features can share the same label, Lin et al. [69] considered individual po-leaders and labels with multiple ports, one for each feature with that label. They proved that minimizing crossings in 1-sided many-to-one boundary labeling with po-leaders is NP-hard, but they also presented a greedy heuristic iteratively assigning label positions with locally fewest leader crossings. Bekos et al. [9] investigated a different labeling style using hyperleaders (recall Section 2.2). Here a po-hyperleader for k sites consists of k vertical p-segments connecting the sites to a single horizontal o-segment, which connects to the label. They considered three problem variants. For crossing-free labelings, the problems are to minimize the number of labels or the total leader length, and for labelings with crossings, the problem is to minimize their number. Both problems with crossing-free leaders can be solved in polynomial time using dynamic programming. Crossing minimization can be solved in $\mathcal{O}(n \cdot c)$ time, where n is the number of sites and c is the number of labels, if the label order is given; if the order is unconstrained, the problem becomes NP-hard.

Gedicke et al. [47] considered dynamic boundary labeling for the 1-sided case. As application they considered small-screen devices such as smart watches. In order to guarantee that all point features are labeled they introduced three different visualization techniques, namely distributing the labels over pages, allowing the user to slide the labels along the bottom side of the screen and stacking the labels on a small number of piles that can be explored by the user; see Figure 6.2. They used po-leaders and considered a multi-criteria objective linearly balancing the number of crossings, the total leader length and the vertical distances between horizon-

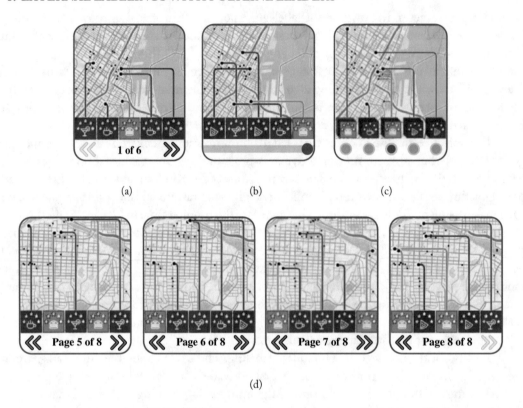

Figure 6.2: Boundary labeling techniques proposed by Gedicke et al. [47]. The labels can be explored either (a) by clicking through pages, (b) by sliding the labels along the boundary, or (c) by clicking through stacks of labels. (d) An example of the last four pages of a multi-page labeling that consists of eight pages in total.

tal leader segments. They considered exact algorithms such as mathematical programming and sweep algorithms as well as heuristics such as hill climbing.

2-Sided Boundary Labeling

A natural extension of the 1-sided case is to admit 2-sided placement of labels, either on two opposite or two adjacent sides of the bounding rectangle. This makes the algorithmic optimization problem more complex, as the distribution of labels to either side becomes an additional degree of freedom.

For uniform labels with fixed positions on two opposite sides of C_A, Bekos et al. [16] presented an $\mathcal{O}(n^2)$-time algorithm for computing a length-minimal solution; see Figure 6.3 for a sample labeling. Using dynamic programming (Section 4.2.1) they obtained a length-minimal assignment of sites and labels, from which crossings were resolved independently on both sides

Figure 6.3: An example of a bend-optimal 2-sided boundary labeling with type-po leaders taken from [16].

as no pair of leaders to opposite sides can intersect in a length-minimal solution. When replacing point features by a certain type of polygonal area features, Bekos et al. [13] gave an $\mathcal{O}(n^2 \log^3 n)$-time algorithm for length-minimal 2-sided boundary labeling, in which the sites of the leaders are located along the sides of the area features. The difference of this algorithm is that the first step to compute the length-minimal site-label assignment uses the weighted matching technique (Section 4.2.2). The crossing removal is done as in the 1-sided case. For uniform labels with sliding positions, Huang et al. [57] presented an $\mathcal{O}(n^5)$-time algorithm as an extension of their algorithm for the 1-sided case. The heuristic framework of Ali et al. [3], already mentioned for 1-sided labeling, can be applied equally well to 2-sided instances.

For instances with obstacles, the dynamic programming algorithms presented by Fink and Suri [39] for minimizing the total length of a labeling with 1-sided po-leaders can be extended to 2-sided instances, at the cost of a significant increase in time complexity from $\mathcal{O}(n^4)$ to $\mathcal{O}(n^9)$ for fixed label positions and from $\mathcal{O}(n^7)$ to $\mathcal{O}(n^{15})$ for sliding label positions.

For arbitrary objective functions, Benkert et al. [18] generalized their dynamic-programming algorithm for 1-sided po-leader labeling to 2-sided boundary labeling, albeit with a worst-case running time of $\mathcal{O}(n^8)$. The reason for the algorithm's much higher complexity compared to $\mathcal{O}(n^2)$ for length minimization [16] is that for non-length-minimal solutions the sites can no longer be easily split into two independent subproblems for the two opposite sides. The extension of Kindermann et al. [62] for non-uniform labels in 1-sided boundary labeling is not properly generalized to optimize 2-sided instances. Instead, for their text labeling application, the authors proposed to simply split the sites at the horizontal median position into two balanced parts and then run the 1-sided algorithm for each side independently.

Many-to-one labeling has also been studied for the 2-sided setting with labels on opposite sides. Since the crossing minimization problem considered by Lin et al. [69] remains NP-hard for the 2-sided case, they again described a heuristic algorithm to obtain a labeling, in which all features sharing the same label connect with their own po-leader to a single label box on one of the two opposite boundaries of C_A. Their algorithm constructs a weighted graph, in which they (heuristically) determine a vertex bisection of minimum weight, which induces the partition of labels to the two boundary sides. The weights are chosen such that the computed bisection has few leaders possibly crossing each other. Then each side is solved with their 1-sided algorithm. The hyperleader labeling by Bekos et al. [9] is also considered in a 2-sided setting, but using duplicated labels that are vertically aligned on both sides. This actually makes the problem algorithmically simpler because now each hyperleader spans the entire width of the image region and splits the instance into two parts. Therefore, the running times of the algorithms for crossing-free labelings with minimum number of labels or minimum length are improved to $\mathcal{O}(n)$ and $\mathcal{O}(n^2)$ running times, respectively. The crossing minimization problem that was shown to be generally NP-hard for the 1-sided case remains open for 2-sided labelings. For a fixed label order, the polynomial-time 1-sided algorithm for crossing minimization can also be applied to the 2-sided case.

A different 2-sided setting is to have labels on two adjacent sides of C_A, e.g., the top and the right side. Here it is no longer the case that a crossing-free solution always exists. Kindermann et al. [63] presented an $\mathcal{O}(n^2)$-time dynamic programming algorithm to compute a valid crossing-free solution if one exists, both for labels with fixed or sliding ports. Further, they presented an algorithm to maximize the number of labeled sites in a crossing-free labeling in $\mathcal{O}(n^3 \log n)$ time. Finally, they modified their algorithm so that it can minimize the total leader length rather than computing some crossing-free solution, however, at the cost of an increase of the time complexity to $\mathcal{O}(n^8 \log n)$ and a space bound of $\mathcal{O}(n^6)$. Bose et al. [22] improved this to an $\mathcal{O}(n^3 \log n)$-time algorithm, again using the dynamic programming technique. An interesting algorithmic open problem is to investigate the remaining difference in the $\mathcal{O}(n^2)$ running time for length minimization with labels on two opposite sides of C_A and the $\mathcal{O}(n^3 \log n)$ running time for labels on two adjacent sides.

Multi-Sided Boundary Labeling

Kindermann et al. [63] studied 3-sided labelings and showed that this problem can be reduced to splitting an instance into two 2-sided subproblems. This results in an $\mathcal{O}(n^4)$-time algorithm working in linear space for finding a crossing-free solution (if one exists) or for maximizing the number of labeled features. For 4-sided instances, they proposed an algorithm that considers all possibilities to split into two special 3-sided instances to be solved as before. This results in an increase of the running time to $\mathcal{O}(n^9)$. Bose et al. [22] improved upon this by presenting new

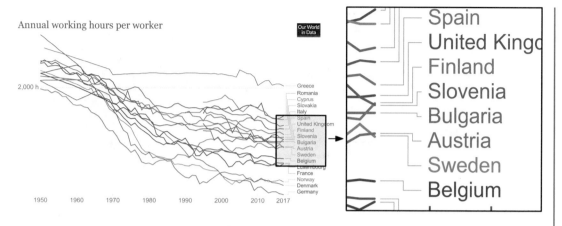

Figure 6.4: Illustration of a sample opo-labeling: average working hours per worker over a full year (plotted in logarithmic scale). The data estimates cover total hours worked in the economy as measured from primarily National Accounts data.

algorithms for 3- and 4-sided boundary labeling that can even find length-minimal or bend-minimal solutions, for fixed and sliding ports and optionally with the leaders avoiding a set of obstacles. This is achieved by reducing the boundary labeling problem to a weighted independent set problem in outerstring graphs, for which a recently published algorithm is used [61]. Recently, Bose et al. [23] further improved this using dynamic programming to achieve $\mathcal{O}(n^3 \log n)$ running time for the 3-sided case and $\mathcal{O}(n^5)$ running time for the 4-sided case. From a heuristic perspective, again the labeling framework of Ali et al. [3] is flexible in terms of the contour C_A and can thus be applied to place labels at three or four sides of a rectangle, as well as around a more complex silhouette contour of the image.

6.2.2   opo-LEADERS

Orthogonal opo-leaders have a slightly more complex shape than the corresponding po-leaders since they contain an additional segment (see Figure 6.4 for a sample illustration). As a result, the readability of the labelings obtained by using such leaders is inferior to the ones obtained by using s-, po-, or do-leaders, as observed by Barth et al. [6]. On the positive side, however, the central p-segment of each opo-leader lies outside the image region R (which, for this particular type of leaders, is usually assumed to be rectangular, as otherwise the readability of the obtained labelings further decreases) routed within a so-called *track routing area* that is wide enough to accommodate all p-segments. This implies that opo-leaders are particularly useful for text annotation purposes [62, 71], since the first o-segment of each opo-leader, which is the only one that interferes with the underlying text, can be drawn between the lines of the text (see Figure 6.5).

Hello, here is some text without a meaning. This text should show what a printed text will look like at this place. If you read this text, you will get no information. Really? Is there no information? Is there a difference between this text and some nonsense like "Huardest gefburn"? Kjift – not at all! A blind text *like this* gives you information about the selected font, how the letters are written and an impression of the look. This text should contain *all letters of the alphabet* and it should be written in of the original language. There is no need for special content, but the length of words should match the language.

Hello, here is some text without a meaning. This text should show what a printed text will look like at this place. If you read this text, you will get no information. Really? Is there no information? Is there a difference between this text and some nonsense like "Huardest gefburn"? Kjift – not at all! A blind text *like this* gives you information about the selected font, how the letters are written and an impression of the look. This text should contain *all letters of the alphabet* and it should be written in of the original language. There is no need for special content, but the length of words should match the language.

Hello, here is some text without a meaning. This text should show what a printed text will look like at this place. If you read this text, you will get no information. Really? Is there no information? Is there a difference between this text and some nonsense like "Huardest gefburn"? Kjift – not at all! A blind text *like this* gives you information about the selected font, how the letters are written and an impression of the look. This text should contain *all letters of the alphabet* and it should be written in of the original language. There is no need for special content, but the length of words should match the language.

Hello, here is some text without a meaning. This text should show what a printed text will look like at this place. If you read this text, you will get no information. Really? Is there no information? Is there a difference between this text and some nonsense like "Huardest gefburn"? Kjift – not at all! A blind text *like this* gives you information about the selected font, how the letters are written and an impression of the look. This text should contain *all letters of the alphabet* and it should be written in of the original language. There is no need for special content, but the length of words should match the language.

Hello, here is some text without a meaning. This text should show what a printed text will look like at this place. If you read this text, you will get no information. Really? Is there no information? Is there a difference between this text and some nonsense like "Huardest gefburn"? Kjift – not at all! A blind text *like this* gives you information about the selected font, how the letters are written and an impression of the look. This text should contain *all letters of the alphabet* and it should be written in of the original language. There is no need for special content, but the length of words should match the language.

First comment

The second comment is a bit longer.

More comments.

short

This comment is much longer. It needs more lines of text than the previous ones.

This is the last comment in this paragraph. The next two paragraphs do not contain any comments.

Comment in another paragraph

Second comment in same line

Third one

Figure 6.5: A document with notes produced by the todonotes package of LaTeX; the visual association of the text and the labels is ambiguous and can be significantly improved.

1-Sided Boundary Labeling

In the 1-sided setting, computing a feasible labeling, in which no two opo-leaders intersect, is an easy task. Assuming without loss of generality that the external labels are attached on a vertical side of the image region R, we observe that the vertical order of the sites must be identical to the vertical order of their corresponding labels. This observation directly gives rise to an $\mathcal{O}(n \log n)$-time constructive algorithm [16]. For the problem of finding labelings, in which the total number

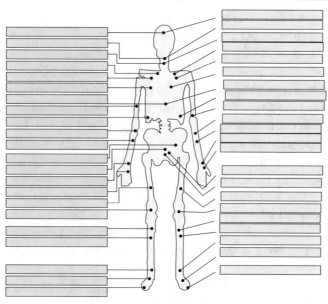

Figure 6.6: An example of a 1-sided boundary labeling of a skeleton. Left side: type-opo leaders created by the approach presented in [15]. Right side: labeling created by hand.

of bends is minimized, Bekos et al. [16] proposed an $\mathcal{O}(n^2)$-time dynamic programming based algorithm to find bend-minimal solutions, assuming that the ports of the labels are sliding; see Figure 6.6 for an illustration. For the case that the ports of the labels are fixed, Huang et al. [57] proposed a slightly less efficient algorithm (which is also based on dynamic programming) to find a bend-minimal solution in $\mathcal{O}(n^3)$ time. In the context of collinear sites, Bekos et al. [14] proposed an $\mathcal{O}(n \log n)$-time algorithm to compute length-minimal solutions by employing a linear-time reduction to a single machine scheduling problem. Notably, their algorithm is not difficult to be adjusted to the more general 1-sided boundary labeling setting assuming that the ports of the labels are sliding; for details refer to Section 4.2.3.

For the case that rectangular-shaped obstacles (e.g., internal labels) are allowed within the image region R, which must not be crossed by the leaders, Fink and Suri [39] presented an $\mathcal{O}(n^{11})$-time algorithm to compute a length-minimal solution, under the assumption that the labels are of uniform size. Bekos et al. [11] and Löffler et al. [73] studied more general mixed labelings, in which each site can be labeled either with an internal label that is located directly to its top-right or with an external label through an opo-leader. When labels are of uniform height and the external labels are to the right side of the image region R, Bekos et al. [11] provided an $\mathcal{O}(n \log n)$-time algorithm for maximizing the total number of internal labels. Note that the case where the external labels are to the left side of R is not symmetric and it turns out to be more difficult. For this case, Bekos et al. [11] provided a quasi-polynomial algorithm of $\mathcal{O}(n^{\log n+3})$

time complexity, a more efficient 2-approximation and an integer linear programming formulation for maximizing the number of internal labels. Löffler et al. [73] presented an improved algorithm to solve the case, in which the external labels are on the left side of R and they are connected to their corresponding sites with type-o leaders, in time $\mathcal{O}(n^3(\log n + \delta))$, where δ denotes the minimum of n and the inverse of the distance of the closest pair of points.

1-sided boundary labelings with opo-leaders have also been studied in the many-to-one setting, in which a point can be attached to more than one label [68, 69]. Lin et al. [69] observed that in this setting crossings will inevitably occur, and proved that minimizing their number is NP-hard. Finally, they presented a 3-approximation algorithm, which places the labels in "median order" as introduced by Eades and Wormald [35]. In a subsequent work, Lin [68] allowed more than one occurrence of the same label, in order to avoid crossings that negatively affect the readability of the produced labelings. Naturally, the focus of his work was on minimizing the total number of used labels; a task which in the 1-sided case can be solved in $\mathcal{O}(n \log n)$ time by a simple top to bottom traversal of the sites, assuming that the labels are along a vertical side of the image region R.

We conclude this subsection by mentioning that 1-sided boundary labeling with opo-leaders has also been studied when more than one stack of labels are allowed along the side s of the image region R containing the labels, under the additional assumption that the labels are all of uniform size and their ports are sliding [12]; see Figure 6.7 for a sample labeling. In this setting, Bekos et al. [12] studied the problem of finding boundary labelings, in which the height of each label is maximized, assuming that the side s is vertical. Polynomial-time algorithms are given for the case of two stacks of labels, and for the case of three stacks of labels under the additional assumption that each opo-leader that connects a point to a label of the second (third) stack must have its p-segment in the gap between the first and the second (second and third, respectively) stack of labels.

1.5-Sided Boundary Labeling

In this setting, the labels are attached along a single side of the image region R, say without loss of generality the right, as in the ordinary 1-sided boundary labeling. However, an opo-leader can be routed to the left side of R temporarily and then finally to the right side, that is, its p-segment is allowed to be on the left side of R. Lin et al. [71] presented an $\mathcal{O}(n^5)$-time algorithm, which—based on dynamic programming—computes a length-minimal solution, under the assumption that the position of the labels are fixed and their sizes are uniform; the ports of the labels can be either fixed or sliding. Lin et al. [70] extended this algorithm to the bend-minimization problem keeping the time complexity unchanged, and they presented an improved algorithm to compute length-minimal solutions in $\mathcal{O}(n \log n)$ time.

Figure 6.7: An example of a multi-stack, 1-sided boundary labeling with type-opo leaders taken from [12].

2-Sided Boundary Labeling

In the 2-sided setting, the algorithmic optimization becomes more complex due to the additional degree of freedom raised by the distribution of the labels to two sides. For uniform labels with either fixed or sliding ports on two opposite sides of the image region R, Bekos et al. [16] presented an $\mathcal{O}(n^2)$-time algorithm for computing a length-minimal solution. Their approach is based on dynamic programming and exploits the fact that in order to avoid crossings, the vertical order of the sites must be identical to the vertical order of their corresponding labels at each of the two sides of R. Under the same set of assumptions, Huang et al. [57] and Fink and Suri [39] proposed less efficient algorithms (which are also based on dynamic programming) to find bend-minimal solutions in $\mathcal{O}(n^5)$ time, and length-minimal solutions in the presence of rectangular-shaped obstacles within the image region R in $\mathcal{O}(n^{27})$ time, respectively. The time complexity of the algorithm by Fink and Suri [39] was improved by Bose et al. [22] to $\mathcal{O}(n^9)$, under the additional assumption that the position of the labels are fixed. Note that if the labels are of non-uniform heights and their positions are not fixed, then both optimization problems mentioned above (i.e., the minimization of the total leader length and the minimization of the total number of leader-bends) become NP-hard, even in the absence of obstacles [16, 57]. They also remain NP-hard in the special case in which the input sites are collinear [14].

In the many-to-one setting, computing a 2-sided boundary labeling with opo-leaders and minimum number of leader crossings is NP-hard, even in the case in which the same number of labels must be attached on two opposite sides of the image region R and the labels are of uniform heights with either fixed or sliding ports [69]; for an illustration refer to Figure 6.8(a).

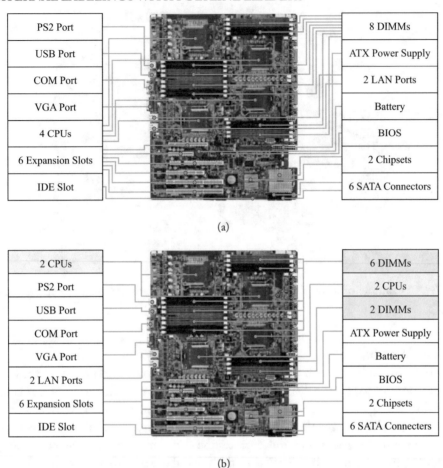

Figure 6.8: Examples of many-to-one 2-sided boundary labelings taken from [69] and [68]. In (a) the leaders are of type opo, while in (b) hyperleader are used (duplicated labels are highlighted in gray).

For this particular setting, Lin et al. [69] suggested an algorithm with an approximation factor of at least three that also depends on the structure of the input instance. If more than one copy of the same label is allowed, Lin [68] provided an $\mathcal{O}(n^2)$-time algorithm which yields a crossing-free routing of the leaders while simultaneously minimizing the number of used labels; for an illustration refer to Figure 6.8(b).

Finally, in the 2-sided setting, in which the labels are given along two adjacent sides of the image region R, Bose et al. [22] gave an $\mathcal{O}(n^{12})$-time algorithm to compute a length-minimal solution (if any), under the assumption that the position of the labels are fixed. Notably, their algorithm works also in the presence of obstacles.

4-Sided Boundary Labeling

In the 4-sided setting, there are significantly fewer results, and they concern boundary labelings of minimal length. To the best of our knowledge, there are no results concerning bend-minimal boundary labelings or boundary labelings that are optimal in terms of some different objective function.

Bekos et al. [16] presented a polynomial-time algorithm to compute length-minimal labelings, assuming that the position of the labels around the image region R are fixed (i.e., specified as part of the input). Their algorithm consists of two steps. In the first step, a (not necessarily crossing-free) length-minimal site-label assignment is computed by finding a minimum-cost perfect matching on an appropriately defined bipartite graph (for details refer to Section 4.2.2). In the second step, the crossings are eliminated such that the total leader length of the assignment computed in the previous step is not affected (i.e., by keeping the solution optimal). The time complexity of the algorithm is dominated by the time needed to compute the minimum-cost perfect matching in the first step and depends on whether the ports of the labels are fixed or sliding. More concretely, in the former case the bipartite graph is geometric, which allows for an efficient minimum-cost perfect matching computation in $\mathcal{O}(n^2 \log^3 n)$ time [92]. In the latter case, however, the bipartite graph has no special property, and therefore a minimum-cost perfect matching must be computed using the Hungarian method, which needs $\mathcal{O}(n^3)$ time; see, e.g., [65].

We conclude this section by mentioning that the aforementioned algorithm was extended to the case of area features by Bekos et al. [13]. In the case of area features with constant number of corners, the computation of the minimum-cost perfect matching, which determines the complexity of the algorithm, is done on a bipartite graph that is not necessarily geometric, which implies that the Hungarian method must be used (as in the case of sliding label ports above), yielding again a time complexity of $\mathcal{O}(n^3)$.

6.2.3   do- AND pd-LEADERS

Orthodiagonal leaders with one bend have a simple shape, which seems not to introduce clutter in the obtained labelings [6]; see, e.g., Figure 6.9 for a sample labeling. In the framework of boundary labeling, in which the image contour C_A is an axis-aligned rectangle, these types of leaders were introduced and first studied by Benkert et al. [18] back in 2009. On the other hand, if C_A is a circle, then early works that adopt the paradigm of do-leaders date back to excentric labeling for interactive labeling of focus regions [37].

Let us first assume that C_A is an axis-aligned rectangle. As already mentioned, Benkert et al. [18] introduced boundary labelings with do-leaders and preliminarily observed that a crossing-free solution might not always be feasible. For a general objective function, they pro-

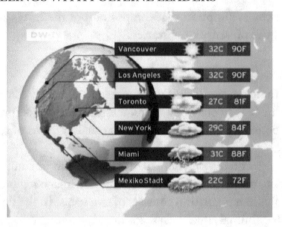

Figure 6.9: A labeling with orthodiagonal leaders used in weather forecast map by DW-TV.

posed a dynamic programming approach, which in $\mathcal{O}(n^5)$ time (in $\mathcal{O}(n^{14})$ time) determines whether there exist an optimal crossing-free boundary labeling with do-leaders, assuming that the ports of the labels are sliding while the positions of the labels are fixed along a single side (along two opposite sides, respectively) of the image region R; in the positive case, the labeling can be reported without increasing the time complexity. A significantly improved algorithm was given, when the objective function is the minimization of the total leader length and the labels are along a single side of R. For this case, the authors suggested a sweep-line algorithm that determines whether there exists an optimal crossing-free boundary labeling with do-leaders in $\mathcal{O}(n^2)$ time.

In order to overcome infeasibility issues, in a subsequent work, Bekos et al. [10] extended the study of boundary labelings with do-leaders by also incorporating od- and pd-leaders, and proved that a combination of these types of leaders guarantees the existence of a crossing-free solution. More precisely, they proved that a crossing-free solution (i.e., that is not optimal under some objective function) with od- and pd-leader can always be derived by a plane-sweep technique in $\mathcal{O}(n \log n)$ time, assuming that the position of the labels are fixed along the image region R. In the same setting, a length-minimal solution can be found in $\mathcal{O}(n^3)$ time by appropriately adjusting both steps of the matching-based algorithm given in [16] (see also Section 6.2.2).

Yang et al. [105] observed that while the algorithm by Bekos et al. [16] can ensure the absence of crossings in a length-minimal solution, it cannot guarantee adequate separation between leaders and connection sites or other leaders, which may lead to ambiguity. Hence, they proposed a quadratic program to further improve the output labelings of the algorithm of Bekos [16] and they used them in a framework, called MapTrix, for visualising many-to-many flows between different geographic locations by connecting an OD matrix [95] with origin and destination maps [100]. Löffler et al. [73] extended the study of boundary labelings with do-leaders in the

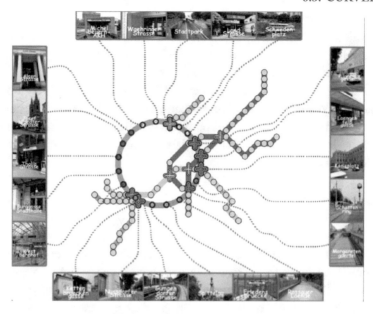

Figure 6.10: An example of 4-sided crossing-free boundary labelings with curved leaders [101].

mixed setting, where also internal labels are allowed, and provided a dynamic programming based algorithm to maximize the number of internal labels.

6.2.4   OTHER POLYLINE LEADERS

Few works studied external labelings with polyline leaders that are neither orthogonal nor orthodiagonal. In text annotation, Kindermann et al. [62] proposed boundary labelings with os-leaders, where the first segment of each leader is used to reach the margin of the text, while the second one has a slope such that all labels can be arranged on the margin of the text. Obviously, such labelings can be derived from opo-labelings by removing the last bend of each leader. Fekete and Plaisant [37] and Bertini et al. [19], who studied excentric labeling, also proposed site-label connections with os-leaders. Here, the first segment of each leader is orthogonal to the boundary of the lens, while the slope of the second segment is chosen such that all labels can be arranged on the boundary of the circular focus region.

6.3   CURVED LEADERS

In this section, we discuss research that uses smooth curves as leaders. In the context of automatically creating metro maps, Wu et al. [103] suggested to use 2-sided boundary labeling and curved leaders based on B-splines to annotate stations with images. They required that the resulting labeling is plane, while also minimizing the number of intersections between leaders and the metro map. At the core of their approach, they used a general flow-network model to construct such leaders. In a user study based on eye-tracking, they showed that the curved leaders outperform straight-line leaders and orthogonal leaders. Wu et al. [101] further extended this approach to 4-sided boundary labeling. To that end, they first partitioned the image region into four regions using the method by Bekos et al. [16] and labeled each region independently with the approach by Wu et al. [103]. Kindermann et al. [62] used Bézier curves to connect text annotations with their sites. Their method first computes a plane straight-line labeling. Afterward, each leader is interpreted as a cubic Bézier curve whose control points are iteratively moved using a force-based method.

6.4   GUIDELINES

Polyline leaders are usually chosen for external labelings for their highly regular shapes, which provide an explicit separation from commonly less regular background images. Using external labels with polyline leaders for instances without a clear center, with irregular content, and with arbitrarily located features thus creates a contrasting layer with a clean and well-structured appearance, much like schematic metro maps bring order into meandering subway networks. In contrast to slope-restricted (say, horizontal or vertical) straight-line leaders, as discussed in Chapter 5, allowing one (or two) bends in a leader gives significantly more flexibility in placing external labels. This advantage, however, comes at the cost of possibly higher cognitive load in tracing leaders with bends and thus there is an increased risk for reading errors, which must be taken into account. We recommend to consider the following aspects when choosing to implement or apply a polyline labeling method and refer to Table 6.1 as a selection aid.

1. Select the most appropriate leader type. For most settings, choosing polylines with a single bend is preferable as it generally guarantees a large number of possible placements, while introducing only a single possibly disturbing bend point. Using po-leaders aligns well with the coordinate axes, the image frame (paper or digital screen), and the printed text. However, do- and pd-leaders are a worthwhile alternative, having a smoother appearance and an easier-to-trace bend angle. Readability experiments [6] showed a slight aesthetic preference of the participants for do-leaders over po-leaders, yet the latter had a slightly better task performance (faster and more accurate). Leaders of type opo are most often used and recommended for external labeling of text documents and other row-based content as they route naturally within the gaps between lines of text. Non-rectangular contours like circles are also suited for polyline leaders with an unrestricted first segment, similarly to radial straight leaders (see Chapter 5).

2. Polyline leaders are particularly interesting for rectangular contours, or also circles. They have been much less studied for other types of hulls/silhouettes, which would require in some cases non-trivial adaptations to the algorithms.

3. Proximity between feature and label is a general principle in external labeling. So minimizing the total and individual lengths of polyline leaders, at least as one major component of the objective function, is highly recommended. There are, however, also other important aspects for the quality of a polyline labeling. For instance, as leader segments are parallel, it is advisable to maintain a minimum distance between two neighboring leaders as to avoid difficulties in visual tracing of dense bundles of parallel lines. Other criteria are usually application dependent and can be considered on demand.

4. Collect any other requirements and properties of the instance at hand, such as: (i) Are the labels of uniform size/height or not? (ii) Can labels be placed on all or just some sides of the contour? (iii) Are the ports and label positions fixed or sliding? Moreover, are the features to be labeled points or is it a special setting for labeling text, areas, or advanced models including mixed labelings, dynamic scenarios, or many-to-one approaches? Depending on the answers to these questions, use Table 6.1 to select an appropriate algorithm.

5. Most existing algorithms in the literature on labeling with polyline leaders compute exact solutions, yet in some cases these algorithms are primarily of theoretical interest with high, but polynomial running times. Table 6.1 lists algorithms that have been implemented by their authors, which may help in choosing algorithms that work well in practice. If no suitable exact algorithm satisfies the performance needs, then selecting one of the heuristic approaches might be preferable.

CHAPTER 7

Conclusions and Outlook

The previous chapters provided an in-depth discussion of the state of the art on external labeling algorithms. Our literature review identified that two mostly independent research communities have studied external labeling problems extensively: the visual computing community as well as the algorithms community.

- In visual computing the vast majority of research considers straight-line leaders and presents heuristics taking a large set of drawing criteria into account. In these practical works the concrete application and deployment of the approaches play a major role.

- In contrast, the algorithms community focuses on proving formal quality guarantees in labeling models limited to fewer drawing criteria. Algorithmic and geometric properties and results are primary, while abstraction makes the concrete application become secondary. In these algorithmic works the results are formal models and exact algorithms, mainly for polyline leaders and boundary labeling.

The observations above raise the question what the two communities can learn from each other and how research efforts can be joined to create synergies. In the following we discuss this question in greater detail and state ten open challenges in external labeling. In these challenges we have not included immediate follow-up problems of individual papers (e.g, improve an algorithm's complexity or extend a method to a broader setting). Rather, these challenges emerge from the discussion that we developed so far. They concern questions that are missing in the existing literature and, according to our experience, need attention.

As is evident from Chapters 5 and 6, a plethora of different labeling approaches have been developed and evaluated in the visual computing community. Still, direct comparisons between different results are missing from the literature, as opposed, e.g., to traditional map labeling, where such comparisons are available both in terms of algorithmic efficiency, but also in terms of the quality of the produced labelings [26]. As a result, the approaches are mostly assessed by discussing labelings of selected example illustrations and 3D models. Hence, even if the approaches consider similar settings (possibly in different applications), one can hardly assess which of the approaches is better in the sense that it yields the more readable or more aesthetically pleasing labeling. Since the evaluations that have been conducted so far are mainly preliminary (see Section 3.3), we deem the systematic and quantitative evaluation of the existing approaches to be one of the most important challenges. In order to obtain sustainable results, this particularly requires to make the results comparable. Especially practitioners will benefit from empirically founded readability results and guidelines when choosing a labeling style and an algorithm for their labeling tasks.

Challenge 7.1 *Systematically evaluate and compare different approaches for external labeling.*

This is where the theoretical works on boundary labeling have their particular strength. They use a common formal structure that is divided into two parts: the labeling model (including all parameters, hard and soft constraints) and the algorithms computing labelings within this particular model. Hence, the formalism admits simple comparisons of the models' features. Further, due to the clear distinction between model and algorithm, an evaluation can be split into two independent parts, i.e., the assessment of the labeling model and the assessment of the labeling algorithm.

Multiple labeling models have been proposed in both communities. As observed in Chapters 5 and 6, in the algorithms community these models are mostly specified by defining a particular leader type and a precise but limited objective function (e.g., the total leader length) that takes into account general drawing principles (e.g., proximity, planarity, and straightness/low detour [85, 97]). On the other hand, in the visual computing community, user studies, expert interviews, and the manual analysis of handmade drawings are common tools to extract and assess various additional drawing criteria, which are combined and balanced using multi-criteria optimization heuristics. As is evident from Section 3.3, in both communities an extensive and systematic comparison of the labeling models is missing. Currently, criteria are often selected by common sense, experience, and following general cognitive and visualization principles and guidelines [86]. While this is reasonable, we believe that establishing a set of quantitative, well-justified quality measures specific to external labeling would be of great help when comparing and choosing between different labeling models. It lends itself to include the already extracted drawing criteria (e.g., short, crossing-free leaders) as a basis for such measures, but we must extend this also toward more global measures such as balance, harmony, and interplay between image and callouts. Introducing new measures requires a systematic justification and ranking.

Challenge 7.2 *Establish and rank quantitative quality and aesthetics measures for evaluating and comparing external labelings.*

Assuming suitable and computable quality measures are established, we can compare different labeling models by computing and evaluating labelings for large sets of input instances. For example, one might aim at labelings that exclusively optimize a single quality measure such that two different labeling models can be compared with respect to this measure. We note that this procedure is independent from the actual labeling algorithm, but only takes the labeling model into account. However, this raises several problems. First of all, we need algorithms that compute mathematically optimal solutions for labeling problems. While for boundary labeling a large variety of such algorithms exist (as evident from Chapter 6, and Tables 5.1 and 6.1), for contour labeling only the algorithm by Niedermann et al. [80] is known. For the general external labeling problem there exists a large number of heuristics that perform well in specific application scenarios, but there is a lack of exact algorithms as evident from Chapter 5. On the other hand, exact algorithms based on established techniques such as dynamic programming or mathematical programming are expected to have high running times and thus limited practical applicability. Hence, the development of alternative algorithmic techniques that perform well in practice is needed. From a theoretical point of view it is also interesting to find lower bounds on the time complexity for both the underlying computational problems and the developed algorithms.

Challenge 7.3   *Develop efficient algorithms for optimally solving external labeling problems that are applicable in practice.*

We deem such algorithms to be helpful both for evaluating labeling models and heuristics, and for producing high-quality labelings used in professional books such as atlases of human anatomy, where somewhat longer computation times can be tolerated. On the other hand, for dynamic settings such as those that appear in the visualization of 3D models, labelings must be computed in real time. Further, complex constraints that improve temporal coherence must be satisfied. For such applications fast heuristics seem to be the first choice. We stress here that the vast majority of the existing heuristics (see Chapter 5) depend highly on the application domain and as a result they are tailored to the specifications of the problem, which they are designed to solve (e.g., [5, 42, 77]). Therefore, it is of importance to develop more generic heuristics of broader scope that can be deployed in several applications.

Challenge 7.4   *Develop more generic, less problem-specific heuristics and algorithms for external labeling.*

Dynamic and interactive settings have been extensively considered by the visual computing community, while the algorithms community has mainly considered static labelings. In relation to Challenge 7.3, a promising research direction is to build up on that knowledge to integrate temporal coherence and user interaction into formal models and to develop algorithms with formal quality guarantees for dynamic settings. The absence of such algorithms is greatly evident from Tables 5.1 and 6.1; in Table 5.1 observe that the intersection of S4 (temporal coherence) and S7.2 (algorithm type: exact) is empty, while in Table 6.1 there is only one entry (i.e., [81]) in row P9.1 (algorithm type: exact) for dynamic external labelings, which focuses on a very restricted case as discussed in Chapter 6.

Challenge 7.5   *Develop efficient exact algorithms for computing dynamic, temporally coherent, and interactive external labelings.*

To obtain comparable and reproducible experimental results in labeling research, it is necessary to establish benchmark sets (which are common in several other fields, e.g., traditional map labeling [96], graph clustering [4], route planning [20], and graph drawing [33]) and make them available for experiments. Depending on the problem setting, these benchmark sets may consist of geographic data (as, e.g., those in [96]), illustrations, maps, or 3D models each annotated with feature sets of varying type, density, and complexity. The need of these benchmarks is even more evident from the fact that several papers in the literature contain sample labelings obtained by implementations of different approaches (refer to S6.1 in Table 5.1 and to P8.2 in Table 5.1) but no systematic/quantitative evaluations.

Challenge 7.6   *Establish general and representative benchmark sets for external labeling.*

Establishing benchmarks will further facilitate the development and the systematic evaluation of (possibly much faster) heuristics, as their performance will be easy to be evaluated by simply comparing the optimal labelings and the results of these heuristics with respect to different quality measures. To simplify the process of such evaluations we suggest to develop a common external labeling framework and interface that allows to easily plug in different models and algorithms. This particularly may motivate researchers to make their source code available to the scientific community (e.g., only few of the implementations reported in Tables 5.1 and 6.1 are publicly available in source code [25, 37] and just few others provide JavaScript implementations or legacy Java applets). We note that the open source library *OpenLL* is currently being developed and provides an API specification for label rendering in 2D and 3D graphics [67]. As the library has its focus on rendering, only simple placement algorithms are provided so far. We deem this library to be one possible starting point for tackling Challenge 7.7.

Challenge 7.7   *Develop a generic framework for models and algorithms on external labeling.*

We pointed out that there is a division between strong practical work using straight-line leaders and a significant amount of theoretical work on polyline leaders. This can be partially explained by the fact that straight-line leaders are the shortest and simplest visual association between sites and labels, while polyline leaders possess complaisant geometric properties that can be exploited for exact algorithms. Nevertheless, it is far from clear which leader type is to be preferred. In a first formal user study on leader readability, Barth et al. [6] showed that po-leaders outperform s-leaders and that opo-leaders lag far behind. However, those experiments were conducted for boundary labeling using blank background images. Hence, in other applications, such as text annotations, opo-leaders seem to be a better choice because they can run between the text lines, while po- or s-leaders pass through the text. Furthermore, s-, po-, do-, and pd-leaders, as well as curved leaders are all frequently found in professional information graphics. They arguably differ in their aesthetic appeal, which is not only influenced by their geometry, but also by the type and style of the background image. The few existing guidelines as described in

Chapter 3 are valuable resources when assessing different options of external labelings, but they do not cover all types of labelings and are primarily based on practical experience [86, 99]. There is a lack of formal empirical studies and readability results to confirm or adapt these guidelines.

Challenge 7.8 *Establish evidence-based guidelines and rules to determine which leader type suits which application and task best.*

Finally, we consider the combination of internal and external labels to be a fruitful research field, as comparatively little work has been done on this topic. From the algorithmic perspective, only few special cases of mixed labeling have been considered so far that mainly combine very simple internal and external labeling models [11, 72, 73]. From a practical, visual perspective, the question that arises is when to use or not to use external rather than internal labels. In the same direction, it is also worth studying more sophisticated mixed labeling models, which allow combinations of internal and external labels, but also support *internal* labels connected to their features by short leaders; see, e.g., [17, 74].

Challenge 7.9 *Develop rules to address the dichotomy between internal and external labels, i.e., when to use which type of labels.*

A natural such rule to address the dichotomy between internal and external labels is to prefer external labeling for features close to the boundary of the background image and internal labeling for features in the core of the background image, as this approach facilitates the visual association of features and their labels. Also, it seems natural to label sparse regions of features with internal labels, while for more dense regions external labels are preferred. Of course, this is also related to the underlying background image, as internal labels may obscure important information underneath. In any case, overlaps between leaders and labels must be avoided to guarantee readability. However, more sophisticated labeling models that yield mixed labelings of high quality are still missing from the literature and need to be established.

Challenge 7.10 *Broaden the research on mixed labeling by combining more sophisticated labeling models for internal and external labels.*

In summary, both in the visual computing community and in the algorithms community a multitude of specializations and variants of external labeling have been considered from two different perspectives. This book provides a comprehensive overview over both fields discussing the research and algorithmic techniques that have been presented over the last two decades. At the same time it shall pave the way to join both research directions creating new synergies and pushing forward the progress on external labeling.

Correction to: Visual Aspects of External Labeling

Correction to:

M.A. Bekos, B. Niedermann and M. Nöllenburg,

Chapter 3 Visual Aspects of External Labeling,
https://doi.org/10.1007/978-3-031-02609-6_3

The Figure 3.3 has incorrectly updated in the chapter 3 of this book and that has now been corrected.

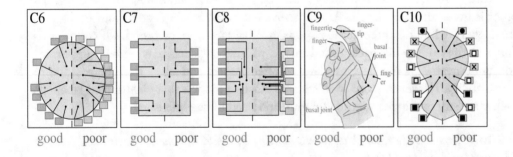

The updated online version of this chapter can be found at
https://doi.org/10.1007/978-3-031-02609-6_3.

Bibliography

[1] P. K. Agarwal, A. Efrat, and M. Sharir. Vertical decomposition of shallow levels in 3-dimensional arrangements and its applications. *SIAM Journal on Computing*, 29:912–953, 1999. doi:10.1137/S0097539795295936. (cited on pp. 41)

[2] R. K. Ahuja, T. L. Magnanti, and J. B. Orlin. *Network flows - theory, algorithms and applications*. Prentice Hall, 1993. (cited on pp. 40)

[3] K. Ali, K. Hartmann, and T. Strothotte. Label layout for interactive 3D illustrations. *Journal of the WSCG*, 13(1):1–8, 2005. URL: http://wscg.zcu.cz/wscg2005/Papers_2005/Journal/!WSCG2005_Journal_Final.pdf. (cited on pp. 4, 25, 29, 33, 52, 53, 55, 58, 59, 65, 70, 71, 73, 76, 78, 81, 83)

[4] D. A. Bader, H. Meyerhenke, P. Sanders, C. Schulz, A. Kappes, and D. Wagner. Benchmarking for graph clustering and partitioning. In *Encyclopedia of Social Network Analysis and Mining*, pages 73–82. Springer, 2014. doi:10.1007/978-1-4614-6170-8_23. (cited on pp. 98)

[5] J. Balata, L. Čmolík, and Z. Mikovec. On the selection of 2D objects using external labeling. In *Human Factors in Computing Systems (CHI'14)*, pages 2255–2258. ACM, 2014. doi:10.1145/2556288.2557288. (cited on pp. 29, 52, 53, 56, 65, 97)

[6] L. Barth, A. Gemsa, B. Niedermann, and M. Nöllenburg. On the readability of leaders in boundary labeling. *Information Visualization*, 2018. doi:10.1177/1473871618799500. (cited on pp. 28, 29, 46, 51, 52, 53, 57, 60, 70, 71, 72, 76, 77, 83, 89, 92, 98)

[7] S. E. Battersby, J. E. Stewart, A. L.-D. Fede, K. C. Remington, and K. Mayfield-Smith. Ring maps for spatial visualization of multivariate epidemiological data. *Journal of Maps*, 7(1):564–572, 2011. doi:10.4113/jom.2011.1182. (cited on pp. 52, 53, 55, 56, 59)

[8] K. Been, E. Daiches, and C. Yap. Dynamic map labeling. *IEEE Transactions on Visualization and Computer Graphics*, 12(5):773–780, Sep. 2006. doi:10.1109/TVCG.2006.136. (cited on pp. 27)

[9] M. Bekos, S. Cornelsen, M. Fink, S.-H. Hong, M. Kaufmann, M. Nöllenburg, I. Rutter, and A. Symvonis. Many-to-one boundary labeling with backbones. *Journal of Graph Algorithms and Applications*, 19(3):779–816, 2015. doi:10.7155/jgaa.00379. (cited on pp. 48, 70, 71, 73, 79, 82)

[10] M. A. Bekos, M. Kaufmann, M. Nöllenburg, and A. Symvonis. Boundary labeling with octilinear leaders. *Algorithmica*, 57(3):436–461, 2010. doi:10.1007/s00453-009-9283-6. (cited on pp. 22, 40, 41, 42, 48, 70, 71, 76, 90)

[11] M. A. Bekos, M. Kaufmann, D. Papadopoulos, and A. Symvonis. Combining traditional map labeling with boundary labeling. In *Current Trends in Theory and Practice of Computer Science (SOFSEM'11)*, volume 6543 of *LNCS*, pages 111–122. Springer, 2011. doi:10.1007/978-3-642-18381-2_9. (cited on pp. 35, 70, 71, 73, 75, 76, 85, 99)

[12] M. A. Bekos, M. Kaufmann, K. Potika, and A. Symvonis. Multi-stack boundary labeling problems. In *Foundations of Software Technology and Theoretical Computer Science (FSTTCS'06)*, volume 4337 of *LNCS*, pages 81–92. Springer, 2006. doi:10.1007/11944836_10. (cited on pp. 48, 70, 71, 73, 79, 86, 87)

[13] M. A. Bekos, M. Kaufmann, K. Potika, and A. Symvonis. Area-feature boundary labeling. *Computer Journal*, 53(6):827–841, 2010. doi:10.1093/comjnl/bxp087. (cited on pp. 40, 41, 70, 71, 81, 89)

[14] M. A. Bekos, M. Kaufmann, and A. Symvonis. Efficient labeling of collinear sites. *Journal of Graph Algorithms and Applications*, 12(3):357–380, 2008. doi:DOI:10.7155/jgaa.00170. (cited on pp. 42, 48, 70, 71, 76, 85, 87)

[15] M. A. Bekos, M. Kaufmann, A. Symvonis, and A. Wolff. Boundary labeling: Models and efficient algorithms for rectangular maps. In *Graph Drawing (GD'04)*, volume 3383 of *LNCS*, pages 49–59. Springer, 2004. doi:10.1007/978-3-540-31843-9_7. (cited on pp. 52, 53, 57, 85)

[16] M. A. Bekos, M. Kaufmann, A. Symvonis, and A. Wolff. Boundary labeling: Models and efficient algorithms for rectangular maps. *Computational Geometry: Theory and Applications*, 36(3):215–236, 2007. doi:10.1016/j.comgeo.2006.05.003. (cited on pp. 4, 40, 41, 44, 47, 55, 57, 61, 70, 71, 76, 77, 80, 81, 84, 85, 87, 89, 90, 92)

[17] B. Bell, S. Feiner, and T. Höllerer. View management for virtual and augmented reality. In *User Interface Software and Technology (UIST'01)*, pages 101–110. ACM, 2001. doi:10.1145/502348.502363. (cited on pp. 99)

[18] M. Benkert, H. J. Haverkort, M. Kroll, and M. Nöllenburg. Algorithms for multi-criteria boundary labeling. *Journal of Graph Algorithms and Applications*, 13(3):289–317, 2009. doi:10.7155/jgaa.00189. (cited on pp. 38, 39, 43, 70, 71, 73, 77, 78, 79, 81, 89)

[19] E. Bertini, M. Rigamonti, and D. Lalanne. Extended excentric labeling. *Computer Graphics Forum*, 28(3):927–934, 2009. doi:10.1111/j.1467-8659.2009.01456.x. (cited on pp. 29, 70, 71, 73, 76, 91)

[20] J. Blum and S. Storandt. Scalability of route planning techniques. In M. de Weerdt, S. Koenig, G. Röger, and M. T. J. Spaan, editors, *Automated Planning and Scheduling (ICAPS'18).*, pages 20–28. AAAI Press, 2018. URL: https://aaai.org/ocs/index.php/ICAPS/ICAPS18/paper/view/17741. (cited on pp. 98)

[21] P. Bobák, L. Čmolík, and M. Čadík. Temporally stable boundary labeling for interactive and non-interactive dynamic scenes. *Computers & Graphics*, 91:265–278, 2020. doi: 10.1016/j.cag.2020.08.005. (cited on pp. 52, 53, 63, 64)

[22] P. Bose, P. Carmi, J. M. Keil, S. Mehrabi, and D. Mondal. Boundary labeling for rectangular diagrams. In *Algorithm Theory (SWAT'18)*, volume 101 of *LIPIcs*, pages 12:1–12:14. Schloss Dagstuhl - Leibniz-Zentrum fuer Informatik, 2018. doi:10.4230/LIPIcs.SWAT.2018.12. (cited on pp. 39, 70, 71, 76, 82, 87, 88)

[23] P. Bose, S. Mehrabi, and D. Mondal. Faster multi-sided one-bend boundary labelling. In R. Uehara, S. Hong, and S. C. Nandy, editors, *WALCOM: Algorithms and Computation (WALCOM'21)*, volume 12635 of *LNCS*, pages 116–128. Springer, 2021. doi:10.1007/978-3-030-68211-8_10. (cited on pp. 70, 71, 83)

[24] P. Brucker. *Scheduling algorithms (4th ed.)*. Springer, 2004. (cited on pp. 42)

[25] S. Bruckner and E. Gröller. Volumeshop: An interactive system for direct volume illustration. In *Visualization (Vis'2005)*, pages 671–678. IEEE, 2005. doi:10.1145/1187112.1187183. (cited on pp. 4, 52, 53, 60, 98)

[26] J. Christensen, J. Marks, and S. M. Shieber. An empirical study of algorithms for point-feature label placement. *ACM Trans. Graph.*, 14(3):203–232, 1995. doi:10.1145/212332.212334. (cited on pp. 95)

[27] L. Čmolík and J. Bittner. Layout-aware optimization for interactive labeling of 3D models. *Computers & Graphics*, 34(4):378–387, 2010. doi:10.1016/j.cag.2010.05.002. (cited on pp. 19, 52, 53, 55, 56, 59)

[28] L. Čmolík and J. Bittner. Real-time external labeling of ghosted views. *IEEE Transactions on Visualization and Computer Graphics*, 2018. doi:10.1109/TVCG.2018.2833479. (cited on pp. 29, 52, 53, 55, 56, 59, 60)

[29] L. Čmolík, V. Pavlovec, H.-Y. Wu, and M. Nöllenburg. Mixed labeling: Integrating internal and external labels. *IEEE Trans. Visualization and Computer Graphics*, 2020. doi:10.1109/TVCG.2020.3027368. (cited on pp. 52, 53, 56, 60)

[30] T. H. Cormen, C. E. Leiserson, R. L. Rivest, and C. Stein. *Introduction to Algorithms (3rd Edition)*. MIT Press, 2009. (cited on pp. 31, 35, 37)

[31] M. de Berg, O. Cheong, M. van Kreveld, and M. H. Overmars. *Computational geometry: algorithms and applications.* Springer, 2008. (cited on pp. 44)

[32] A. Degani. On the typography of flight-deck documentation. Technical Report 177605, Ames Research Center, 1992. (cited on pp. 24)

[33] G. Di Battista and W. Didimo. Gdtoolkit. In R. Tamassia, editor, *Handbook on Graph Drawing and Visualization.*, pages 571–597. Chapman and Hall/CRC, 2013. (cited on pp. 98)

[34] P. Eades. A heuristic for graph drawing. *Congressus numerantium*, 42:146–160, 1984. (cited on pp. 33)

[35] P. Eades and N. C. Wormald. Edge crossings in drawings of bipartite graphs. *Algorithmica*, 11(4):379–403, 1994. doi:10.1007/BF01187020. (cited on pp. 48, 86)

[36] N. Elmqvist, A. V. Moere, H.-C. Jetter, D. Cernea, H. Reiterer, and T. Jankun-Kelly. Fluid interaction for information visualization. *Information Visualization*, 10(4):327–340, 2011. doi:10.1177/1473871611413180. (cited on pp. 27)

[37] J.-D. Fekete and C. Plaisant. Excentric labeling: Dynamic neighborhood labeling for data visualization. In *Human Factors in Computing Systems (CHI'99)*, pages 512–519. ACM Press, 1999. doi:10.1145/302979.303148. (cited on pp. 4, 29, 52, 53, 56, 59, 62, 70, 71, 73, 76, 89, 91, 98)

[38] M. Fink, J.-H. Haunert, A. Schulz, J. Spoerhase, and A. Wolff. Algorithms for labeling focus regions. *IEEE Transactions on Visualization and Computer Graphics*, 18(12):2583–2592, 2012. doi:10.1109/TVCG.2012.193. (cited on pp. 52, 53, 55, 56, 59)

[39] M. Fink and S. Suri. Boundary labeling with obstacles. In *Canadian Conference on Computational Geometry (CCCG'16)*, pages 86–92, 2016. (cited on pp. 48, 52, 53, 55, 57, 70, 71, 78, 81, 85, 87)

[40] M. Formann and F. Wagner. A packing problem with applications to lettering of maps. In *Computational Geometry (SoCG'91)*, pages 281–288. ACM Press, 1991. doi:10.1145/109648.109680. (cited on pp. 27)

[41] T. M. J. Fruchterman and E. M. Reingold. Graph drawing by force-directed placement. *Software: Practice and experience*, 21(11):1129–1164, 1991. doi:10.1002/spe.4380211102. (cited on pp. 33)

[42] G. Fuchs, M. Luboschik, K. Hartmann, K. Ali, H. Schumann, and T. Strothotte. Adaptive labeling for interactive mobile information systems. In *Information Visualization (InfoVis'06)*, pages 453–459. IEEE, 2006. doi:10.1109/IV.2006.17. (cited on pp. 52, 53, 62, 97)

[43] J. L. Gabbard, J. E. Swan, II, D. Hix, R. S. Schulman, J. Lucas, and D. Gupta. An empirical user-based study of text drawing styles and outdoor background textures for augmented reality. In *Proc. of the IEEE Conference on Virtual Reality*, pages 11–18, 2005. doi:10.1109/VR.2005.7. (cited on pp. 27)

[44] M. R. Garey and D. S. Johnson. *Computers and Intractability: A Guide to the Theory of NP-Completeness*. W. H. Freeman, 1979. (cited on pp. 47)

[45] M. R. Garey, R. E. Tarjan, and G. T. Wilfong. One-processor scheduling with symmetric earliness and tardiness penalties. *Mathematics of Operations Research*, 13(2):330–348, 1988. doi:10.1287/moor.13.2.330. (cited on pp. 48)

[46] M. A. Garrido, C. Iturriaga, A. Márquez, J. R. Portillo, P. Reyes, and A. Wolff. Labeling subway lines. In *Algorithms and Computation (ISAAC'01)*, volume 2223 of *LNCS*, pages 649–659. Springer, 2001. doi:10.1007/3-540-45678-3_55. (cited on pp. 48)

[47] S. Gedicke, A. Bonerath, B. Niedermann, and J.-H. Haunert. Zoomless maps: External labeling methods for the interactive exploration of dense point sets at a fixed map scale. *IEEE Transactions on Visualization and Computer Graphics*, 27:1247–1256, 2021. doi:10.1109/TVCG.2020.3030399. (cited on pp. 70, 71, 79, 80)

[48] A. Gemsa, J.-H. Haunert, and M. Nöllenburg. Multirow boundary-labeling algorithms for panorama images. *ACM Transactions on Spatial Algorithms and Systems*, 1(1):1–30, 2015. doi:10.1145/2794299. (cited on pp. 42, 48, 52, 53, 55, 56, 62, 64)

[49] T. Götzelmann, K. Ali, K. Hartmann, and T. Strothotte. Form follows function: Aesthetic interactive labels. In *Computational Aesthetics in Graphics, Visualization and Imaging (CAe'05)*, pages 193–200. Eurographics Association, 2005. doi:10.2312/COMPAESTH/COMPAESTH05/193-200. (cited on pp. 59)

[50] T. Götzelmann, K. Hartmann, and T. Strothotte. Agent-based annotation of interactive 3D visualizations. In *Smart Graphics (SG'06)*, volume 4073 of *LNCS*, pages 24–35. Springer, 2006. doi:10.1007/11795018_3. (cited on pp. 52, 53, 61)

[51] T. Götzelmann, K. Hartmann, and T. Strothotte. Contextual grouping of labels. In *SimVis*, pages 245–258, 2006. URL: http://www.simvis.org/Tagung2006/sv-proceedings.html. (cited on pp. 52, 53, 55, 56, 62)

[52] R. Grasset, M. Tatzgern, T. Langlotz, D. Kalkofen, and D. Schmalstieg. Image-driven view management for augmented reality browsers. In *Proc. IEEE International Symposium on Mixed and Augmented Reality (ISMAR) 2012*, pages 177–186, Atlanta GA, USA, November 2012. doi:10.1109/ISMAR.2012.6402555. (cited on pp. 24, 27)

[53] K. Hartmann, K. Ali, and T. Strothotte. Floating labels: Applying dynamic potential fields for label layout. In *Smart Graphics (SG'04)*, volume 3031 of *LNCS*, pages 101–113. Springer, 2004. `doi:10.1007/978-3-540-24678-7_10`. (cited on pp. 29, 33, 52, 53, 54, 63, 64)

[54] K. Hartmann, T. Götzelmann, K. Ali, and T. Strothotte. Metrics for functional and aesthetic label layouts. In *Smart Graphics (SG'05)*, volume 3638 of *LNCS*, pages 115–126. Springer, 2005. `doi:10.1007/11536482_10`. (cited on pp. 29, 52, 53, 59, 62)

[55] J.-H. Haunert and T. Hermes. Labeling circular focus regions based on a tractable case of maximum weight independent set of rectangles. In *ACM SIGSPATIAL Workshop on MapInteraction*, 2014. `doi:10.1145/2677068.2677069`. (cited on pp. 18, 52, 53, 59, 60)

[56] N. Heinsohn, A. Gerasch, and M. Kaufmann. Boundary labeling methods for dynamic focus regions. In *Pacific Visualization Symposium (PacificVis'14)*, pages 243–247. IEEE, 2014. `doi:10.1109/PacificVis.2014.20`. (cited on pp. 52, 53, 55, 59, 62)

[57] Z. Huang, S. Poon, and C. Lin. Boundary labeling with flexible label positions. In *WALCOM: Algorithms and Computation (WALCOM'14)*, volume 8344 of *LNCS*, pages 44–55. Springer, 2014. `doi:10.1007/978-3-319-04657-0_7`. (cited on pp. 22, 42, 48, 70, 71, 74, 76, 78, 81, 85, 87)

[58] E. Imhof. Positioning names on maps. *The American Cartographer*, 2(2):128–144, 1975. (cited on pp. 27)

[59] S. Julier, Y. Baillot, D. G. Brown, and M. Lanzagorta. Information filtering for mobile augmented reality. *IEEE Computer Graphics and Applications*, 22(5):12–15, 2002. `doi:10.1109/MCG.2002.1028721`. (cited on pp. 27)

[60] M. Kaufmann. On map labeling with leaders. In S. Albers, H. Alt, and S. Näher, editors, *Efficient Algorithms*, volume 5760 of *LNCS*, pages 290–304. Springer, 2009. `doi:10.1007/978-3-642-03456-5_20`. (cited on pp. 69)

[61] M. Keil, J. Mitchell, D. Pradhan, and M. Vatshelle. An algorithm for the maximum weight independent set problem on outerstring graphs. *Computational Geometry: Theory and Applications*, 60:19–25, 2017. `doi:10.1016/j.comgeo.2016.05.001`. (cited on pp. 83)

[62] P. Kindermann, F. Lipp, and A. Wolff. Luatodonotes: Boundary labeling for annotations in texts. In *Graph Drawing (GD'14)*, volume 8871 of *LNCS*, pages 76–88. Springer, 2014. `doi:10.1007/978-3-662-45803-7_7`. (cited on pp. 52, 53, 55, 56, 57, 58, 70, 71, 74, 79, 81, 83, 91, 92)

[63] P. Kindermann, B. Niedermann, I. Rutter, M. Schaefer, A. Schulz, and A. Wolff. Multi-sided boundary labeling. *Algorithmica*, 76(1):225–258, 2016. `doi:10.1007/s00453-015-0028-4`. (cited on pp. 39, 70, 71, 73, 82)

[64] C. Koulamas. Single-machine scheduling with time windows and earliness/tardiness penalties. *European Journal of Operational Research*, 91(1):190–202, 1996. `doi:https://doi.org/10.1016/0377-2217(95)00116-6`. (cited on pp. 42)

[65] H. W. Kuhn. The Hungarian method for the assignment problem. *Naval Research Logistic Quarterly*, 2:83–97, 1955. (cited on pp. 40, 89)

[66] A. Leykin and M. Tuceryan. Determining text readability over textured backgrounds in augmented reality systems. In *Proc. of the ACM SIGGRAPH international conference on Virtual Reality continuum and its applications in industry*, pages 436–439, 2004. `doi:10.1145/1044588.1044683`. (cited on pp. 27)

[67] D. Limberger, A. Gropler, S. Buschmann, B. Wasty, and J. Döllner. Openll: An api for dynamic 2d and 3d labeling. In *Information Visualisation (IV'18)*, pages 175–181, July 2018. `doi:10.1109/iV.2018.00039`. (cited on pp. 98)

[68] C. Lin. Crossing-free many-to-one boundary labeling with hyperleaders. In *Pacific Visualization Symposium (PacificVis'10)*, pages 185–192. IEEE, 2010. `doi:10.1109/PACIFICVIS.2010.5429592`. (cited on pp. 34, 70, 71, 73, 76, 86, 88)

[69] C. Lin, H. Kao, and H. Yen. Many-to-one boundary labeling. *J. Graph Algorithms Appl.*, 12(3):319–356, 2008. `doi:10.7155/jgaa.00169`. (cited on pp. 35, 48, 70, 71, 73, 76, 79, 82, 86, 87, 88)

[70] C. Lin, S. Poon, S. Takahashi, H. Wu, and H. Yen. One-and-a-half-side boundary labeling. In *Combinatorial Optimization and Applications (COCA'11)*, volume 6831 of *LNCS*, pages 387–398. Springer, 2011. `doi:10.1007/978-3-642-22616-8_30`. (cited on pp. 22, 42, 48, 70, 71, 74, 76, 86)

[71] C. Lin, H. Wu, and H. Yen. Boundary labeling in text annotation. In *Information Visualization (InfoVis'09)*, pages 110–115. IEEE, 2009. `doi:10.1109/IV.2009.84`. (cited on pp. 35, 70, 71, 73, 74, 76, 83, 86)

[72] M. Löffler and M. Nöllenburg. Shooting bricks with orthogonal laser beams: A first step towards internal/external map labeling. In *Canadian Conference on Computational Geometry (CCCG'10)*, pages 203–206, 2010. URL: http://cccg.ca/proceedings/2010/paper54.pdf. (cited on pp. 70, 71, 73, 75, 79, 99)

[73] M. Löffler, M. Nöllenburg, and F. Staals. Mixed map labeling. *J. Spatial Information Science*, 13:3–32, 2016. `doi:10.5311/JOSIS.2016.13.264`. (cited on pp. 70, 71, 73, 75, 85, 86, 90, 99)

[74] M. Luboschik, H. Schumann, and H. Cords. Particle-based labeling: Fast point-feature labeling without obscuring other visual features. *IEEE Trans. Vis. Comput. Graph.*, 14(6):1237–1244, 2008. `doi:10.1109/TVCG.2008.152`. (cited on pp. 99)

[75] J. B. Madsen, M. Tatzqern, C. B. Madsen, D. Schmalstieg, and D. Kalkofen. Temporal coherence strategies for augmented reality labeling. *IEEE Transactions on Visualization and Computer Graphics*, 22(4):1415–1423, 2016. `doi:10.1109/TVCG.2016.2518318`. (cited on pp. 27, 29)

[76] S. Masuda, K. Nakajima, T. Kashiwabara, and T. Fujisawa. Crossing minimization in linear embeddings of graphs. *IEEE Transactions on Computers*, 39(1):124–127, 1990. `doi:10.1109/12.46286`. (cited on pp. 48)

[77] K. Mogalle, C. Tietjen, G. Soza, and B. Preim. Constrained labeling of 2D slice data for reading images in radiology. In *Visual Computing for Biomedicine (VCBM'12)*, pages 131–138. Eurographics Association, 2012. `doi:10.2312/VCBM/VCBM12/131-138`. (cited on pp. 25, 29, 52, 53, 55, 56, 65, 97)

[78] K. Mühler and B. Preim. Automatic textual annotation for surgical planning. In *Vision, Modeling, and Visualization (VMV'09)*, pages 277–284. Eurographics Association, 2009. (cited on pp. 29, 52, 53, 56, 65)

[79] B. Niedermann and J. Haunert. Focus+context map labeling with optimized clutter reduction. *International Journal of Cartography*, 5(2–3):158–177, 2019. Special issue of 29th International Cartographic Conference (ICC'19). `doi:10.1080/23729333.2019.1613072`. (cited on pp. 3, 52, 53, 65, 66)

[80] B. Niedermann, M. Nöllenburg, and I. Rutter. Radial contour labeling with straight leaders. In *Pacific Visualization Symposium (PacificVis'17)*. IEEE, 2017. `doi:10.1109/PACIFICVIS.2017.8031608`. (cited on pp. 4, 25, 28, 29, 51, 52, 53, 54, 55, 56, 61, 97)

[81] M. Nöllenburg, V. Polishchuk, and M. Sysikaski. Dynamic one-sided boundary labeling. In *Advances in Geographic Information Systems (ACM-GIS'10)*, pages 310–319. ACM Press, 2010. `doi:10.1145/1869790.1869834`. (cited on pp. 70, 71, 74, 78, 97)

[82] S. Oeltze-Jafra and B. Preim. Survey of labeling techniques in medical visualizations. In *Visual Computing for Biomedicine (VCBM'14)*, pages 199–208. Eurographics Association, 2014. `doi:10.2312/vcbm.20141192`. (cited on pp. 51)

[83] S. D. Peterson, M. Axholt, and S. R. Ellis. Objective and subjective assessment of stereoscopically separated labels in augmented reality. *Computers & Graphics*, 33(1):23–33, 2009. `doi:10.1016/j.cag.2008.11.006`. (cited on pp. 27)

[84] S. Pick, B. Hentschel, I. Tedjo-Palczynski, M. Wolter, and T. Kuhlen. Automated Positioning of Annotations in Immersive Virtual Environments. In T. Kuhlen, S. Coquillart, and V. Interrante, editors, *Joint Virtual Reality Conference of EGVE (JVRC'09)*. The Eurographics Association, 2010. `doi:10.2312/EGVE/JVRC10/001-008`. (cited on pp. 27, 29, 52, 53)

[85] H. Purchase. Which aesthetic has the greatest effect on human understanding? In *Graph Drawing (GD'97)*, volume 1353 of *LNCS*, pages 248–261. Springer, 1997. `doi:10.1007/3-540-63938-1_67`. (cited on pp. 25, 26, 27, 96)

[86] C. Richards. Technical and scientific illustration. In A. Black, P. Luna, O. Lund, and S. Walker, editors, *Information Design: Research and Practice*, chapter 5, pages 85–106. Taylor & Francis, 2017. `doi:10.4324/9781315585680`. (cited on pp. 23, 96, 99)

[87] M. Shimabukuro and C. Collins. Abbreviating text labels on demand. In *Information Visualization (INFOVIS'17)*. IEEE, 2017. Poster abstract. (cited on pp. 25)

[88] T. Stein and X. Décoret. Dynamic label placement for improved interactive exploration. In *Non-Photorealistic Animation and Rendering (NPAR'08)*, pages 15–21. ACM Press, 2008. `doi:10.1145/1377980.1377986`. (cited on pp. 52, 53, 56, 65)

[89] M. Tatzgern, D. Kalkofen, R. Grasset, and D. Schmalstieg. Hedgehog labeling: View management techniques for external labels in 3D space. In *Virtual Reality (VR'14)*, pages 27–32. IEEE, 2014. `doi:10.1109/VR.2014.6802046`. (cited on pp. 19, 27, 52, 53)

[90] M. Tatzgern, D. Kalkofen, and D. Schmalstieg. Dynamic compact visualizations for augmented reality. In *Virtual Reality (VR'13)*, pages 3–6. IEEE, 2013. `doi:10.1109/VR.2013.6549347`. (cited on pp. 52, 53, 55, 56, 64)

[91] E. R. Tufte. *The Visual Display of Quantitative Information*. Graphics Press, 2001. (cited on pp. 25, 27)

[92] P. M. Vaidya. Geometry helps in matching. *SIAM Journal on Computing*, 18:1201–1225, 1989. `doi:10.1137/0218080`. (cited on pp. 41, 89)

[93] R. J. Vanderbei. *Linear Programming: Foundations and Extensions*. Springer US, 2014. (cited on pp. 45)

[94] I. Vollick, D. Vogel, M. Agrawala, and A. Hertzmann. Specifying label layout style by example. In *User Interface Software and Technology (UIST'07)*, pages 221–230. ACM Press, 2007. `doi:10.1145/1294211.1294252`. (cited on pp. 29, 34, 52, 53, 55, 62)

[95] A. M. Voorhees. A general theory of traffic movement. *Transportation*, 40(6):1105–1116, 2013. `doi:10.1007/s11116-013-9487-0`. (cited on pp. 90)

[96] F. Wagner and A. Wolff. A practical map labeling algorithm. *Comput. Geom.*, 7:387–404, 1997. doi:10.1016/S0925-7721(96)00007-7. (cited on pp. 98)

[97] C. Ware, H. Purchase, L. Colpoys, and M. McGill. Cognitive measurements of graph aesthetics. *Information Visualization*, 1:103–110, 2002. doi:10.1057/palgrave.ivs.9500013. (cited on pp. 27, 96)

[98] A. Wolff. Graph drawing and cartography. In R. Tamassia, editor, *Handbook of Graph Drawing and Visualization*, chapter 23, pages 697–736. CRC Press, 2013. (cited on pp. 4)

[99] C. H. Wood. A descriptive and illustrated guide for type placement on small scale maps. *Cartographic J.*, 37(1):5–18, 2000. doi:10.1179/caj.2000.37.1.5. (cited on pp. 99)

[100] J. Wood and J. Dykes. Spatially ordered treemaps. *IEEE Transactions on Visualization and Computer Graphics*, 14(6):1348–1355, 2008. doi:10.1109/TVCG.2008.165. (cited on pp. 90)

[101] H.-Y. Wu, S.-H. Poon, S. Takahashi, M. Arikawa, C.-C. Lin, and H.-C. Yen. Designing and annotating metro maps with loop lines. In *Information Visualization*, pages 9–14. IEEE, 2015. doi:10.1109/iV.2015.14. (cited on pp. 91, 92)

[102] H.-Y. Wu, S. Takahashi, C.-C. Lin, and H.-C. Yen. A zone-based approach for placing annotation labels on metro maps. In *Smart Graphics (SG'11)*, volume 6815 of *LNCS*, pages 91–102. Springer, 2011. doi:10.1007/978-3-642-22571-0_8. (cited on pp. 35, 52, 53, 62, 63)

[103] H.-Y. Wu, S. Takahashi, C.-C. Lin, and H.-C. Yen. Travel-route-centered metro map layout and annotation. In *Computer Graphics Forum*, volume 31, pages 925–934. Wiley Online Library, 2012. doi:10.1111/j.1467-8659.2012.03085.x. (cited on pp. 23, 29, 91, 92)

[104] W. Wu and E. N. Dalal. Perception-based line quality measurement. In *Image Quality and System Performance II*, volume 5668. SPIE, 2005. doi:10.1117/12.585968. (cited on pp. 24)

[105] Y. Yang, T. Dwyer, S. Goodwin, and K. Marriott. Many-to-many geographically-embedded flow visualisation: An evaluation. *IEEE Transactions on Visualization and Computer Graphics*, 23(1):411–420, 2017. doi:10.1109/TVCG.2016.2598885. (cited on pp. 29, 70, 71, 73, 76, 90)

Authors' Biographies

MICHAEL A. BEKOS

Michael A. Bekos is currently a postdoctoral researcher at the Algorithmic group of the Department of Computer Science of the University of Tübingen in Germany, while recently he was appointed as an assistant professor at the Department of Mathematics of the University of Ioannina in Greece. In 2009, he received a Ph.D. in Theoretical Computer Science from the National Technical University of Athens. His research interests primarily focus on the development of algorithms to solve problems mostly from the research areas of Map Labeling, Graph Drawing, and Graph Theory. His research work counts more than 100 peer-reviewed research papers, and among them more than 15 on various aspects of labeling.

BENJAMIN NIEDERMANN

Benjamin Niedermann obtained his Ph.D. in Computer Science from the Karlsruhe Institute of Technology (KIT), Germany, in 2017. From 2017–2021, he was a member of the research group Geoinformation at the University of Bonn. His research interests comprise the development of efficient algorithms in Computational Geometry, Computational Cartography, and Geoprocessing. One main focus of his research is label placement in figures, maps, and dynamic scenes. It includes mathematical models, the design of algorithms with provable guarantees, as well as the empirical evaluation of the algorithms in real-world scenarios.

MARTIN NÖLLENBURG

Martin Nöllenburg is a full professor in the Algorithms and Complexity Group at TU Wien, Vienna, Austria. He obtained a Ph.D. and a habilitation degree in Computer Science from the Karlsruhe Institute of Technology (KIT) in 2009 and 2015, respectively, and joined TU Wien as an assistant professor in 2015. His research interests include the engineering of graph and geometric algorithms, in particular for the visualization of networks and spatial data. He has published more than 140 peer-reviewed research papers, and among them more than 35 on various aspects of internal and external labeling algorithms.

Index

do-leader, 37–41, 83, 89, 90

opo-leader, 17, 21, 41–43, 83–87, 91

pd-leader, 17, 41, 90

po-leader, 17, 39, 44, 48, 72, 79, 83

1-sided boundary labeling, 16, 21, 42–44, 57, 58, 73, 76–79, 84–86

1.5-sided boundary labeling, 86

2-sided boundary labeling, 16, 41, 57, 73, 80–82, 87, 88, 91

2-sided boundary labeling with adjacent sides, 16, 82

3-sided boundary labeling, 16, 73, 82, 83

4-sided boundary labeling, 16, 40, 44, 82, 83, 89, 92

anchor, 12

background layer, 11, 15, 16, 20, 23, 26, 28

boundary labeling, 4, 15, 17, 22, 35, 39–44, 46–48, 54, 57, 73, 77, 79, 83, 89–91, 95, 96, 98

callout labeling, 2, 14, 20, 23, 31, 32, 37, 38, 46, 55, 59, 61, 65, 73

contour labeling, 4, 15, 16, 21, 34, 65, 73, 74, 77, 83

crossing-free labeling, 14, 21, 33, 35, 40, 43, 44, 55, 62, 65, 89, 90

curved leaders, 17, 23, 91, 92

drawing area, 11, 19, 33, 36–38

dynamic labeling, 19, 27, 29, 55, 56, 64, 69–72, 79

dynamic programming, 35–37, 39, 56, 57, 59, 61, 62, 76, 78–83, 85–87, 90, 91

excentric labeling, 4, 16, 21, 29, 47, 59, 62, 65, 73, 89, 91

feasible labeling, 20, 21, 25, 35, 73, 74, 84, 89

feature layer, 11

fixed port, 13, 20, 21, 38, 40, 41, 43, 44, 74, 76–78, 82, 83, 85–87, 89

focus-region labeling, 4, 16, 47, 62, 72, 89, 91

force-based approach, 33, 34, 56, 62, 64, 65, 78, 92

greedy algorithm, 31, 32, 56, 59–62, 64, 65, 76, 79

hyperleader, 6, 18, 79, 82

image region, 11, 13, 15, 19, 21, 42, 44, 47, 51, 62, 64, 65, 72, 73, 75, 77, 82–90

labeling layer, 11, 23, 26

labeling region, 11

leader, 1, 12, 17

lens, 4, 16, 18, 32, 62, 65

many-to-one labeling, 6, 18, 48, 69–73, 76, 79, 82, 86, 87

mathematical programming, 45, 46, 63, 76, 80

multi-stack boundary labeling, 48, 79

orthodiagonal leader, 17, 72, 77, 89, 91

orthogonal leader, 17, 77, 92

plane labeling, 14, 39, 40, 57, 59, 91, 92

plane sweep, 43, 44, 90

port, 13, 14, 20, 21, 44, 74, 75, 82

radial labeling, 16, 32

scheduling, 42, 43, 48, 74, 76, 85

sliding port, 13, 41, 43, 74, 78, 79, 82, 83, 85–87, 89, 90

straight-line leader, 17, 23, 24, 32, 51, 56–58, 62

weighted matching, 40, 41, 56, 76, 78, 81, 89, 90

Printed in the United States
by Baker & Taylor Publisher Services